D. A. Smith

Annals of the CEREMADE

Edited by
J.P. Aubin
A. Bensoussan
I. Ekeland

Birkhäuser
Boston · Basel · Stuttgart

Advances in Hamiltonian Systems

J.P. Aubin,
A. Bensoussan,
I. Ekeland,
editors

1983

Birkhäuser
Boston • Basel • Stuttgart

Editors:

J.P. Aubin
C.E.R.E.M.A.D.E.
Université de Paris IX Dauphine
F-75775, Paris Cedex 16
FRANCE

A. Bensoussan
C.E.R.E.M.A.D.E.
Université de Paris IX Dauphine
F-75775, Paris Cedex 16
FRANCE

I. Ekeland
C.E.R.E.M.A.D.E.
Université de Paris IX
F-75775, Paris Cedex 16
FRANCE

Library of Congress Cataloging in Publication Data
Main entry under title:

Advances in Hamiltonian systems.

(Annals of the CEREMADE)
English and French.
Papers from a conference held at the University of Rome, Feb. 1981 and sponsored by CEREMADE and the Istituto matematico Guido Castelnuovo.
Bibliography: p.
Contents: Recent advances in the study of the existence of periodic orbits of Hamiltonian systems / by Antonio Ambrosetti — The direct method in the study of periodic solutions of Hamiltonian systems with prescribed period / by V. Benci — Periodic solutions of Hamiltonian systems having prescribed minimal period / by Giovanni Mancini — [etc.]
1. Hamiltonian systems—Congresses. I. Aubin, Jean Pierre. II. Bensoussan, Alain. III. Ekeland, I. (Ivar), 1944- IV. C.E.R.E.M.A.D.E. V. Istituto matematico Guido Castelnuovo. IV. Series: Annals of the C.E.R.E.M.A.D.E.
QA614.83.A38 1983 514'.74 83-5021
ISBN 3-7643-3130-5 (Switzerland)

CIP-Kurztitelaufnahme der Deutschen Bibliothek

Advances in Hamiltonian systems / J. P. Aubin...ed.
- Boston ; Basel ; Stuttgart : Birkhäuser, 1983.
(Annals of the CEREMADE ; 2)
ISBN 3-7643-3130-5

NE: Aubin, Jean-Pierre [Hrsg.]; Centre de Recherche de Mathématiques de la Décision PARIS: Annals of the...

ISBN 3-7643-3130-5
Printed in USA

CONTENTS

Antonio AMBROSETTI - International School for Advanced Studies
(SISSA)
Strada Costiera 11
Trieste, Italy 34014.

V. BENCI - Instituto di Matematica Applicata.
Universita di Bari.
Italy.

Giovanni MANCINI - Universita di Trieste.
Italy.

Ivar EKELAND - CEREMADE - Université PARIS IX-DAUPHINE ,
Place de Lattre de Tassigny,
75016 PARIS.

Jean-Michel LASRY - CEREMADE - Université PARIS IX-DAUPHINE ,
Place de Lattre de Tassigny,
75016 PARIS.

Pierre BERNHARD - CEREMADE - Université PARIS IX-DAUPHINE ,
Place de Lattre de Tassigny,
75016 PARIS.

Joël BLOT - CEREMADE - Université PARIS IX-DAUPHINE ,
Place de Lattre de Tassigny,
75016 PARIS.

Erick GAUSSENS - CEREMADE - Université PARIS IX-DAUPHINE ,
Place de Lattre de Tassigny,
75016 PARIS.

FOREWORD

These are the proceedings of a conference held in February 1981 at the University of Rome. This meeting was sponsored by the CEREMADE and the Instituto Matematica Guido Castelnuovo, under a standing cooperation agreement signed by the universities of Rome and Paris-Dauphine. We thank both institutions for their support.

The subject of periodic solutions of Hamiltonian systems has been particularly fruitful in the last years. There is a beautiful interplay of methods from ordinary differential equations, nonlinear functional analysis, and geometry, which results in deceiptively simple statements.

The study of periodic solutions of Hamiltonian systems goes back to Poincaré and Liapounov. They studied periodic solutions near a reference solution, or near an equilibrium, by perturbation methods which are still in favour today. The main result in this direction is the celebrated Kolmogorov-Arnold-Moser theorem on invariant tori.

All these results are local, in the sense that they describe situations which are close to a trivial one. They all rely on more or less refined versions of the inverse function theorem. In contrast, the results that can be achieved by modern methods are global in nature. Periodic solutions will be found in the large, without the help of some small parameter. It is hoped that such results will lead to a better understanding of the underlying dynamical system.

The first paper, by A. Ambrosetti, is a survey of these recent, global, results. It describes the dual variational method which has been very successful in the case of convex Hamiltonians. The next paper, by V. Benci, describes the direct variational method, relying on the least action principle. Both papers use the S^1-invariance inherent to an autonomous system.

The third paper, by G. Mancini, is concerned with minimal periods. If a T-periodic solution is found, the question can be raised whether it is not in fact $\frac{T}{2}$ or $\frac{T}{100}$ - periodic, i.e. whether T is its true, or minimal, period. This is a natural question, but difficult to answer, and Mancini surveys all known results in this direction.

The fourth paper, by I. Ekeland and J.M. Lasry, shows that the dual action principle for Hamiltonian systems is but a particular case of a general non-convex duality theory, with applications to other situations. The next paper, by P. Bernhard, deals with Hamiltonian systems arising from optimal control, and shows how to extend to this case the classical theory of the second variation in the calculus of variations.

The next paper, by J. Blot, connects up these modern methods with the classical perturbation expansions. By a straightforward application of the inverse function theorem to the dual action functional, he gets asymptotic expansions in terms of a small parameter. Finally, the last paper, by E. Gaussens, describes an efficient numerical method, based on the recent theoretical advances, for finding periodic solutions.

RECENT ADVANCES IN THE STUDY OF THE EXISTENCE OF PERIODIC ORBITS OF HAMILTONIAN SYSTEMS

Antonio AMBROSETTI

International School for Advanced Studies (SISSA)
Strada Costiera 11
Trieste, Italy 34014.

§ 1. INTRODUCTION

In this paper we shall discuss some recent advances in the study of Hamiltonian systems. In view of other expositions in these Proceedings, we shall limit ourselves to considering the existence of periodic solutions on a prescribed energy surface.

Let $H \in C^1(\mathbb{R}^{2n},\mathbb{R})$; an autonomous Hamiltonian system (with n degrees of freedom) is a system of ordinary differential equations like

$$\begin{cases} \dot{p}_i = -H_{q_i}(p_1,\ldots,p_n,q_1,\ldots,q_n) \\ \dot{q}_i = H_{p_i}(p_1,\ldots,p_n,q_1,\ldots,q_n) \end{cases} \qquad i = 1,\ldots,n$$

where $\dot{p}_i = dp_i/dt$, $H_{p_i} = \partial H/\partial p_i$, etc.

Introducing $z = (p_1,\ldots,p_n,q_1,\ldots,q_n)$, $J = \begin{pmatrix} 0 & -I \\ I & 0 \end{pmatrix}$ with I the identity in $\mathbb{R}^n$, and denoting by H' the gradient of H, the above system can be written as

$$(1) \qquad -J\dot{z} = H'(z)$$

One of the main problems concerning (1) is to find a periodic solution $z = z(t)$ of (1) lying on a prescribed energy surface $H(z) = c$. Note that such a problem makes sense because $H(z(t))$ is constant for every (periodic) solution $z(t)$ of (1).

The first result concerning "small" oscillations, namely such that $H(z) = \varepsilon$, $\varepsilon > 0$ small enough, was the celebrated Liapunov Center

Theorem [1]. More recently Weinstein [2] and Moser [3] extended the Liapunov result eliminating the non-resonance condition :

0. Theorem. Let H be C^2 in a neighborhood of $z = 0$ and suppose the Hessian $H''(0)$ be positive definite. Then for every $\varepsilon > 0$ small enough, (1) has on $H(z) = \varepsilon$ at least n distinct periodic orbits.

These results are local in nature ; they are somewhat related to the "Hopf bifurcation" of periodic solutions ([4 - 7] and references therein). However we do not go into details on the above local problem, our purpose being to discuss more closely the existence of periodic solutions of (1) on an energy surface "in the large".

In section 2 below we shall state some of the main results in this direction. The rest of the paper is devoted to outlining the proofs, which involve several different aspects of the calculus of variations in the large, like : min-max methods, Lusternik-Schnirelman critical points theory, the S^1-genus, etc. It is not exaggerated to say that in the study of Hamiltonian systems all the power of such fascinating variations methods applies.

§ 2. THE THEOREMS

Throughout the paper we shall use the following notations : if $x,y \in \mathbb{R}^M$, $x \cdot y$ denotes the Euclidean scalar product, $|x|^2 = x \cdot x$ and $B_R = \{x \in \mathbb{R}^{2n} : |x| < r\}$.

We shall always assume (unless explicitely stated) that $H \in C'(\mathbb{R}^{2n}, \mathbb{R})$. Let $\Sigma = \{x \in \mathbb{R}^{2n} : H(x) = 1\}$, we look for periodic solutions $z = z(t)$ of (1) such that $z \in \Sigma$. The first existence results (for general Hamiltonian systems) was established by Weinstein [8] who proved :

1. Theorem. Let Ω be a bounded convex domain in $\mathbb{R}^{2n}$ such that $\Sigma = \partial\Omega$ and suppose $H'(z) \neq 0$, $\forall z \in \Sigma$. Then (1) has a periodic orbit on Σ .

As we shall see in section 3, the proof of Theorem 1 can be carried out substituting, first of all, the system (1) with a new one, which has the same trajectories on Σ , but is simpler to study.

Using the same device, but with a different functional framework, Rabinowitz [9] weakened the convexity assumption on Ω , taking Ω starshaped with respect to some point $x_o \in \mathbb{R}^{2n}$. He obtained the following result :

2. Theorem. Suppose that $H'(z) \neq 0$, $\forall z \in \Sigma$ and that Σ is radially diffeomorphic to $\partial B = \{x \in \mathbb{R}^{2n} : |x| = 1\}$. Then (1) has a periodic orbit on Σ .

Lastly we state a remarkable theorem of Ekeland and Lasry [10], which deals the existence of n modes of vibration on Σ . This is the only known result concerning multiple periodic solutions of (1) on a prescribed energy surface, namely a complete extension in the large of Theorem 0 above.

3. *Theorem.* *Let* $\Sigma = \partial\Omega$, *with* *bounded, convex domain in* $\mathbb{R}^{2n}$. *Suppose there exist* r , $R > 0$, *with*

$$R^2 < 2\, r^2 ,$$

such that $B_r \subset \Omega \subset B_R$. *Then (1) has at least n different periodic orbits on* Σ .

All the above theorems are proved by looking for the 2π-periodic solutions of suitable related Hamiltonian systems, but in the case of Theorem 3 it is needed to show there are n of such solutions having 2π as *minimal* period. Very little is known about this latter problem, namely the existence of solutions of (1) with a prescribed minimal period, the only papers on this being one by Clarke-Ekeland [11] when H is subquadratic and [12] by Mancini and the present author when H is superquadratic. See also the paper [13] by Mancini in these Proceedings. We believe that improvements of Theorem 3 are closely related to better results concerning the latter problem.

§ 3. PROOF OF THEOREM 1

The proof of Theorem 1 is carried out in 2 steps : (i) the original problem is substituted by a new one having the same periodic orbits on Σ ; (ii) this problem is reduced to the search of the non-trivial critical points of a "dual" functional ; (iii) such critical points are found as min-max using a well-known result by Rabinowitz and the present author [14].

(i) Let $K \in C^1(\mathbb{R}^{2n},\mathbb{R})$ be such that $\forall\ z \in \Sigma$; it results $K(z) = 1$ and $K'(z) \neq 0$. It is easy to see that the Hamiltonian system

$$(2) \qquad -J\dot{w} = K'(w)$$

has on Σ the same periodic orbits as (1). In fact $\forall\ z \in \Sigma$, $H'(z) = \rho(z)\ K'(z)$, for some positive, continuous $\rho : \Sigma \to \mathbb{R}$; then, if w(t) is a periodic solution of (2) with $w \in \Sigma$, there is a reparametrization s(t), satisfying

$$\frac{ds}{dt} = \rho(w(s)) \qquad s(0) = 0$$

such that $z(t) \equiv w(s(t))$ is a solution of (1) with $z \in \Sigma$. For more details see [15] .

In the present case, we can choose K in the following way : for all $z \in \mathbb{R}^{2n}$, $z \neq 0$, there exists a unique $\bar{z} \in \Sigma$ and a positive $a = a(z)$ such that $z = a\bar{z}$. Setting $K(z) = a^4$, and $K(0) = 0$ (but we could have taken $K(z) = a^{\beta}$, for any $\beta > 2$), we consider the corresponding system (2). Let $\bar{w} = \bar{w}(t)$ be a 2π-periodic solution of (2) and let $k = K(\bar{w}(t))$. In view of the homogeneity of K, one directly verifies that

$$z = k^{-1/4}\,\bar{w}(k^{-1/2}t)$$

is a periodic solution of (2) satisfying $z \in \Sigma$. Hence in order to find the periodic orbits of (1) on Σ , it is enough to find the 2π-periodic solutions of (2), with the homogeneous K as above.

(ii) We are going to discuss a variational principle which enables us to find such 2π-periodic solutions of (2). One possibility would be to look for the stationary points on the Hilbert space of those 2π-periodic functions $z \in L^2(0,2\pi;\mathbb{R}^{2n})$, having first derivatives $\dot{z} \in L^2(0,2\pi;\mathbb{R}^{2n})$, of the functional defined by

$$(3) \qquad -\frac{1}{2}\int z\,J\dot{z} - \int K(z) \quad .$$

(Here and throughout the following, the integrals are evaluated from 0 to 2π). Even if strongly indefinite, (3) has been studied either by a Galerkin argument [9] or directly [16] by a generalization of the saddle-point theorem of [14] (see also theorem 4 below). Here, following papers by Clarke [17] and Ekeland [18], we prefer to present a procedure, the "dual action principle", which requires a convex Hamiltonian but leads to a much simpler functional.

Let $E = \left\{u \in L^{4/3}(0,2\pi;\mathbb{R}^{2n}) \; : \; \int u = 0\right\}$, with norm

$$\|u\| = \left|\frac{1}{2}\int |u|^{4/3}\right|^{3/4} .$$

The densely defined operator $z \to -J\dot{z}$ is invertible on E with compact selfadjoint inverse L . Moreover, since Σ is convex, K is convex as well and then it makes sense to take the Legendre transform G of K, defined by

$$G(u) = \sup\,\{u\cdot z - H(z) \; : \; z \in \mathbb{R}^{2n}\}$$

Note that, since K is homogeneous of degree 4, then G turns out to be well defined, of class $C^1(\mathbb{R}^{2n} \setminus \{0\} , \mathbb{R})$ and homogeneous of degree 4/3. Recall also that G' is actually the inverse of K', hence it results

$$(4) \qquad G'(u) = z \quad \text{iff} \quad K'(z) = u$$

Let us consider the functional $f \in C^1(E,\mathbb{R})$ defined by

$$f(u) = -\frac{1}{2}\int u \cdot Lu + \int G(u)$$

and let $\bar{v} \in E$, $\bar{v} \neq 0$ be a stationary point of f on E. From

$$-\int u \cdot L\bar{v} + \int G'(\bar{v}) \cdot u = 0 \qquad \forall\, u \in E$$

it follows that there exists $\xi \in \mathbb{R}^{2n}$ such that

$$(5) \qquad -L\bar{v} + G'(\bar{v}) = \xi$$

Let

$$z = L\bar{v} + \xi \quad ,$$

we claim that z is a solution of (2). In fact, it results $-J\dot{z} = \bar{v}$; on the other hand, (5) gives $z = G'(\bar{v})$ and hence, in view of (4), we deduce that $-J\dot{z} = \bar{v} = K'(z)$, as claimed. Note that z is non-constant provided $\bar{v} \neq 0$.

(iii) We shall find non-trivial critical points of f as saddle points by using a well-known theorem. Our functional f is unbounded both from below and from above (as that in (3)) but now it is the term $\int G(u)$ which dominates at $u = 0$, because $\int G(u)$ behaves like $\|u\|^{4/3}$, while $-\frac{1}{2}\int u \cdot Lu$ is bilinear. More precisely the following holds :

(I) there exists $\rho, \varepsilon > 0$ such that $f(u) > 0$ for all $0 < \|u\| < \varepsilon$ and $f(u) \geqslant \rho$ for all $\|u\| = \varepsilon$.

Next, noting that $\int u \cdot Lu$ is positive somewhere, we easily verify that :

(II) there exists $\bar{u} \neq 0$ such that $f(\bar{u}) = 0$.

Lastly, it is possible to show (see [18] for more details) that f satisfies the so-called Palais-Smale condition, namely :

(P-S) if $u_n \in E$ is such that $f(u_n)$ is bounded and $f'(u_n) \to 0$ then u_n has a converging subsequence .

In view of (I), (II) and (P-S), we can apply the following Theorem [14, Thm. 2.1] :

4. Theorem. *Let* f *be any* C^1 *functional defined on the Banach space* E . *Suppose* f *verifies (I), (II) and (P-S). Let*

$$P = \{p \in C([0,1], E) : p(0) = 0 , p(1) = \bar{u}\}$$

and

$$b = \inf_{p \in P} \left\{ \max_{\tau \in [0,1]} f(p(\tau)) \right\} .$$

Then $0 < \rho < b < +\infty$, *and there exists* $v \in E$ *such that* $f(v) = b$, $f'(v) = 0$. *In particular*, v *is a non-trivial critical point of* f.

Let us remark that in this dual setting we sketched above, the indeterminacy of $\int u \cdot Lu$ has no consequences on the study of the functional.

This is no longer true in the case of Theorem 2. In fact, in this latter case we can only perform step (i) as before, but the functional defined in (3) does not satisfy hypothesis (I). However, this difficulty can be overcome, for example, by an improvement of Theorem 4 (cf. [16, Thm. 0.1]).

§ 4. LUSTERNIK-SCHNIRELMAN CRITICAL POINTS THEORY IN THE PRESENCE OF S^1-SYMMETRIES.

In order to prove Theorem 3 we have to exploit more deeply the symmetries of our problem, namely the fact that equation (1) (or (2)) is invariant under the S^1 action $u(t) \to u(t+s)$. Our purpose below is to discuss a Ltusternik-Schnirelman critical point theory in this case.

The main topological tool in such a theory is an "index" (or genus) which permits the classification of the invariant sets. We first recall some definitions for the reader's convenience.

For $u \in E$, we have

$$u = \sum_{-\infty}^{\infty} u_k e^{ikt}$$

with $u_k \in C^n$, $u_o = 0$ and $u_{-k} = \bar{u}_k$ (here and in the following $\bar{z}$ denotes the conjugate of $z \in C^n$). Let $u \in E$ and $s \in [0,2\pi)$; we define $A_s \in L(E)$ by

$$A_s u(t) = u(t+s)$$

and denote by A the class of all such A_s , $s \in [0,2\pi)$. A set $X \subset E$ is A-invariant if $A_s u \in X$ for all $u \in X$, $A_s \in A$; a functional $f \in C(E,\mathbb{R})$ is A-invariant if $f(A_s u) = f(u)$ for all $u \in E$, $A_s \in A$; a mapping $\psi \in C(E,E)$ is A-equivariant if $\psi \circ A_s = A_s \circ \psi$ for all $A_s \in A$. For example, our functional

$$f(u) = -\frac{1}{2} \int u \cdot Lu + \int G(u)$$

is clearly $\mathcal{A}$-invariant. Notice that, since for $u \in E$ it results $u_o = 0$ then $\mathcal{A}$ has no non-trivial fixed points, namely $A_s u = u$ for all $A_s \in \mathcal{A}$ iff $u = 0$.

Denoting by Γ the class of those closed subsets of $E \setminus \{0\}$ which are $\mathcal{A}$-invariant, we define a mapping $\gamma : \Gamma \to \mathbb{N} \cup \{+\infty\}$ by the following : $\gamma(X)$ is the smallest integer k such that there exists a mapping $\phi \in C(X, C^k \setminus \{0\})$ and $m \in \mathbb{N}^+$ satisfying

(6) $$\phi(A_s u) = e^{ims} \phi(u)$$

for all $u \in X$, $A_s \in \mathcal{A}$. We set $\gamma(X) = \infty$ if there are no such integers and $\gamma(\emptyset) = 0$.

Some of the properties of γ are listed in the following Lemma :

5. *Lemma*. *Let* $X, Y \in \Gamma$. *Then* :

(i) $X \subset Y$ *implies* $\gamma(X) \leqslant \gamma(Y)$;

(ii) $\gamma(X \cup Y) \leqslant \gamma(X) + \gamma(Y)$;

(iii) *if* $\psi \in C(X,Y)$ is $\mathcal{A}$-equivariant, then $\gamma(Y) \geqslant \gamma(X)$;

(iv) *if X is compact*, $\gamma(X) < \infty$ and *there exists* $\delta > 0$ *such that* $\gamma(X) = \gamma(N_\delta(X))$, *where* $N_\delta(X)$ *denotes the* δ*-neighborhood of X*.

We will also need :

6. *Lemma*. *Let* S^n *denote the following* $\mathcal{A}$*-invariant set in E* :

$$S^n = \{u = (x \cos t + y \sin t , x \sin t - y \cos t) : x,y \in \mathbb{R}^n , |x|^2 + |y|^2 = 1\} .$$

Then $\gamma(S^n) = n$.

For the proof of these two Lemmas, see [19] . Note that in Lemma 5(iv), $\gamma(X) < \infty$ because A has no fixed points $u \neq 0$.

In view of later applications, we consider an A-invariant $h \in C^1(E,\mathbb{R})$ and set $M = \{u \in E , u \neq 0 : h(u) = 0\}$. Suppose M is a smooth manifold in E and look for the critical points of f constrained on M , namely the points $u \in M$ such that $f'(u) = \lambda h'(u)$ for some $\lambda \in \mathbb{R}$. The usual way to proceed in the Lusternik-Schnirelman critical points theory is the following : for every $j \leqslant \gamma(M)$ (remark that $M \in \Gamma$) define M_j to be the class of all compact $X \in \Gamma$, $X \subset M$ such that $\gamma(X) \geqslant j$; suppose $M_j \neq \emptyset$, $\forall j \leqslant \gamma(M)$ and set

$$(7) \qquad c_j = \inf_{X \in M_j} \left\{ \sup_{u \in X} f(u) \right\} .$$

It results in $c_1 \leqslant c_2 \leqslant c_3 \leqslant \dots$ and $c_j < \infty$; moreover, if f is bounded from below on M , one has also $c_1 > -\infty$. Suppose, in addition, that the pair (f,M) satisfies the (P-S) condition :

(P-S) every sequence $u_n \in M$ such that $f(u_n)$ is bounded and $f'_{/M}(u_n) \to 0$, has a converging subsequence.

Then it is possible to show that every c_j is a critical level, namely that there exists at least an $u \in M$, such that $f(u) = c_j$ and $f'_{/M}(u) = 0$.

This relies on the following argument.

Suppose c_j is not a critical level for f on M . Then there exists an $\varepsilon > 0$ and $\psi \in c(M,M)$ such that

$$(8) \qquad \sup_{u \in X} f(\psi(u)) < c_j - \varepsilon$$

for all $X \subset M$ with $\sup \{f(u) : u \in X\} < c_j + \varepsilon$. Let us remark that to define such ψ one uses a "steepest descent method" substituting f' with a "pseudo-gradient vector fields" [27] and taking into account that M is diffeomorphic to a C^2 manifold in E . Moreover, since f and h are assumed to be A-invariant, ψ can be taken A-invariant, too.

Then, according to the definition (7) of c_j, we take $X \in M_j$ such that $\sup\{f(u) : u \in X\} < c_j + \varepsilon$. Using now Lemma 5(iii) we deduce $\gamma(\psi(X)) \geq \gamma(X) \geq j$, so that $\psi(X) \in M_j$. This, jointly with (8), is in contradiction with respect to (7). A more careful analysis, using also Lemma 5(iv), permits the proof of multiplicity results : if $c_\ell = c_{\ell+1} = \dots c_{\ell+r}$ then $f_{/M}$ has , at level c_ℓ , a set K of critical points such that $\gamma(K) \geq r + 1$. In particular, the following result holds :

<u>7 . Theorem</u>. <u>Let</u> $f,h \in C^1(E,\mathbb{R})$ <u>be A-invariant</u>. <u>Suppose</u> $h'(u) \neq 0$ $\forall\, u \in M$ <u>and that there is an A-equivariant diffeomorphism between</u> M <u>and a</u> C^2 <u>manifold in</u> E . <u>Moreover, assume</u> f <u>is bounded from below on</u> M, <u>the pair</u> (f,M) <u>satisfies (P-S) and let</u> $M_j \neq \emptyset$. <u>Then</u> :

(i) <u>every</u> c_j <u>is a critical level for</u> $f|_M$;

(ii) <u>if</u> $c_\ell = \dots = c_{\ell+r}$ <u>then</u> $\gamma(K) \geq r+1$.

Theorem 7 requires a little more care. If u is a critical point of f on M , then every $v \in \mathcal{O}(u) := \{A_s u : s \in [0,2\pi)\}$ is such a critical point, too. Evidently, every point of the orbit $\mathcal{O}(u)$ of u is not "geometrically" distinct from u . Hence we are actually interested in the number of distinct critical orbits of f on M . If $c_j \neq c_\ell$ then the corresponding critical points, say u_j and u_ℓ , have distinct orbits, because $f_{/\mathcal{O}(u_j)} \neq f_{/\mathcal{O}(u_\ell)}$. In the case (ii), when $c_\ell = \dots = c_{\ell+r}$, again f has at least r + 1 critical points u_i , with $\mathcal{O}(u_i)$ different, $i = 1,\dots,r+1$. In fact, since A has no non-trivial fixed points, one has $\gamma(\mathcal{O}(u)) = 1$ for all $u \in E$, $u \neq 0$. Then, if $K = \mathcal{O}(u_1) \cup \dots \cup \mathcal{O}(u_k)$ with $k \leq r$, it follows $\gamma(K) \leq k \leq r$, in contradiction to (ii). In conclusion, to theorem 7 can be added the following remark :

<u>8 . Remark</u>. (i) if $c_j \neq c_\ell$ there the corresponding critical points u_j, u_ℓ are such that $\mathcal{O}(u_j) \neq \mathcal{O}(u_\ell)$; (ii) if $c_\ell = \dots = c_{\ell+r}$ then at level c_ℓ there are at least r+1 critical points $u_{\ell+k}$, $k = 0,\dots,r$, such that $\mathcal{O}(u_{\ell+k})$ are mutually distinct.

§ 5. PROOF OF THEOREM 3

We shall sketch here a proof of Theorem 3, following closely a forthcoming joint paper with Mancini [20]. First of all, we prefer to modify slightly the setting of section 3. To show that the arguments we shall develop do not depend on the choice of K , (cf. section 3, step (i)), we take now $K(z) = a^\beta$, $\beta > 2$ and arbitrary for the remainder. The relationship between a 2π-periodic solution $\bar{w}$ of (2) and the corresponding $z \in \Sigma$ is now :

$$(9) \qquad z = k^{-1/\beta}\, \bar{w}\left(k^{2-\beta/\beta}\, t\right) \qquad , \qquad k = K(\bar{w})$$

In step (ii) we shall take $E = \left\{u \in L^\alpha(0,2;\mathbb{R}^{2n}) : \int u = 0\right\}$ where $\frac{1}{\alpha} + \frac{1}{\beta} = 1$. Analogously the Legendre transform G of K will be α-homogeneous.

Before going on, some remarks are in order. Let z_1, z_2 be two (non-constant) periodic solutions of (1) and suppose there exists a diffeomorphism $\phi(t)$ such that $z_2(t) = z_1(\phi(t))$. From $\dot{z}_2 = \phi'\dot{z}_1$ and $\dot{z}_i = JH'(z_i)$, $i = 1,2$, it follows $\phi'(t) = 1$ and hence $z_2(t) = z_1(t+s)$ for some s . Then, for the corresponding critical points v_1, v_2 of f it results in $v_2 \in \mathcal{O}(v_1)$. Hence, for each pair of critical points of f v_1, v_2 such that $\mathcal{O}(v_1)$ does not coincide with $\mathcal{O}(v_2)$, the corresponding periodic solutions of (1) cannot be obtained one from the other through a reparametrization. But, even if $\mathcal{O}(v_1) \neq \mathcal{O}(v_2)$, these critical points could give rise to the same orbit of (1) on Σ if their <u>minimal</u> period is not 2π . On the contrary, if $v_1 \neq v_2$ (more precisely : if $\mathcal{O}(v_1) \neq \mathcal{O}(v_2)$) and their <u>minimal</u> period os 2 , there the corresponding w_1 and w_2 have the same property. Let $k_i = K(w_i)$. By (9) it follows that the orbits described by z_1 and z_2 are different. In fact either $k_1 = k_2$ or not. In the former case the claim follows because $w_1 \neq w_2$; in the latter because z_i has minimal period $2\pi\, k_i^{(\beta-2)/\beta}$.

Hence we shall look for critical points of f having 2π as minimal period. To this purpose we shall substitute the search of critical points of f on E with an equivalent problem with constraint.

Precisely, let

$$h(u) = \langle f'(u),u\rangle = -\int u\cdot Lu + \int G'(u)\cdot u$$

$$= -\int u\cdot Lu + \alpha \int G(u)$$

and set $M = \{u \in E\ ,\ u \neq 0 : h(u) = 0\}$. For u M it follows that $\langle h'(u),u\rangle = -2\int u\cdot Lu + \alpha^2 \int G(u) = \alpha(\alpha-2)\int G(u)$. Hence $\langle h'(u),u\rangle \neq 0 \quad \forall\, u \in M$. In particular M is a smooth manifold in E . Moreover, if $v \neq 0$ is a critical point of f contrained on M , it results

$$(10) \qquad f'(v) = \lambda\, h'(v)$$

for some $\lambda \in \mathbb{R}$. From (10) it follows :

$$(11) \qquad \langle f'(v),v\rangle = \lambda \langle h'(v),v\rangle$$

The left-hand side of (11) is equal to $h(v) = 0$. On the right-hand side $\langle h'(v),v\rangle \neq 0$ and thus $\lambda = 0$. In other words, $v \neq 0$ is a critical point of f on E if and only if v is a critical point of f on M. This device has been introduced by Nehari [21] and is usually employed to find the critical points of unbounded functionals (cf. [22-23-24], etc.). Actually such results can be obtained in a direct manner (cf. [25,14]). The new feature, here, is to use this device to find critical points having additional properties. For other applications see [12] and [26].

Now, f on M takes the form

$$(12) \qquad f(u) = \frac{2-\alpha}{\alpha} \int G(u) = \frac{2-\alpha}{2\alpha} \int u \cdot Lu \qquad (u \in M)$$

Note also that, given $w \in E$, $w \neq 0$, $\lambda w \in M$ if and only if

$$(13) \qquad \lambda^{2-\alpha} \int w \cdot Lw = \alpha \int G(w)$$

In particular there is an A-equivariant map $\psi : S^n \to M$ defined by $\psi(w) = \lambda w$ with λ given by (13). We shall use this ψ below. Here it suffices to remark that M_n (see the notation introduced in Section 4) is not empty, because $\psi(S^n) \in M_n$. In fact $\psi(S^n)$ is compact and $\gamma(\psi(S^n)) \geqslant \gamma(S^n) = n$ in view of Lemma 5(iii) and Lemma 6.

It is easy to see (cf. [12, Lemma 2.4]) that (f,M) satisfies (P-S). Moreover, from (12) it follows directly that f is bounded from below on M . Hence we are in position to apply the arguments of Section 4. Our goal here will be to show that the assumption $R^2 < 2r^2$ in Theorem 3 ensures that the levels $c_1,\ldots,c_n$ correspond to critical points with minimal period 2π .

First of all, since (f,M) satisfies (P-S), then f attains its minimum on M . Let us denote by m such a minimum ($m = c_1$) and let $m = f(\bar{v})$. Let $M^\star$ be the set of points $u \in M$ such that

$$u = \sum_{-\infty}^{\infty} u_{hk} \, e^{ihkt}$$

for some integer $H \geqslant 2$. The following Lemma holds :

9. <u>Lemma</u>. <u>Let</u> $\delta = \frac{1}{2-\alpha}$. <u>Then</u> $\min\{f(u) : u \in M^\star\} = f(2^\delta \bar{v}) = 2^{\alpha\delta} m$.

The proof of Lemma 9 is carried out by using (12) and (13) ; for more details cf. [20, Lemma 1].

Let $m^\star = \min\{f(u) : u \in M^\star\}$ and $\hat{M} = \{u \in M : f(u) < m^\star\}$ it is clear that every critical point v of $f_{/M}$ such that $v \in \hat{M}$, will have 2π as minimal period. In order to find n distinct critical points of $f_{/M}$ lying in $\tilde{M}$ we first prove :

10. Lemma. Under the assumptions of Theorem 3, it follows that $\psi(S^n) \subset \tilde{M}$.

Assuming the Lemma 10, the proof of Theorem 3 can be easily completed. In fact, recalling that $\psi(S^n) \subset M_n$, it follows that

$$c_n = \inf_{X \in M_n} \left\{ \sup_X f(u) \right\} < \sup_{\psi(S^n)} f(u) < m^\star .$$

Hence the result follows because Theorem 7, Remark 8 and as was discussed before.

It remains to prove Lemma 10. First we remark that the quadratic form $\frac{1}{2}\int u \cdot Lu$ takes its maximum on the unit sphere $S = \{u \in E : \|u\| = 1\}$. Let b denote such maximum, and let $b = \frac{1}{2}\int \bar{u}\cdot L\bar{u}$. It is easy to see that $\bar{u}$ has the form $(\bar{x}\cos t + \bar{y}\sin t , \bar{x}\sin t - \bar{y}\cos t)$ for some $\bar{x},\bar{y} \in \mathbb{R}^N$, $|\bar{x}|^2 + |\bar{y}|^2 = 1$, or, in other words, $\bar{u} \in S^n$.

Proof of Lemma 10. According to the definition of K , it results : $K(x) = |x|^\beta K\left(\frac{x}{|x|}\right)$ (for $|x| \neq 0$) and $K\left(\frac{x}{|x|}\right) = \frac{1}{\rho^\beta}$ where $\rho \in \mathbb{R}^+$ is such that $\rho \frac{x}{|x|} \in \Sigma$. Since $B_r \subset \Omega \subset B_R$, $\Sigma = \partial\Omega$, one has

$$R^{-\beta} |x|^\beta \leqslant K(x) \leqslant r^{-\beta} |x|^\beta$$

From this we deduce the following estimate for G :

$$(14) \qquad G(y) = \sup_{X \in R^{2n}} \{x\cdot y - K(x))\} \leqslant \sup_X \left\{x\cdot y - \frac{1}{R^\beta}|x|^\beta\right\} =$$

$$= a_1 R^\alpha |y|^\alpha$$

where $a_1 = \frac{1}{\alpha} \beta^{1/1-\beta}$.

In the same way :

(15) $$G(y) \geqslant a_1 r^\alpha |y|^\alpha$$

with the same a_1 as in (14).

From (14) and (15) we get

(16) $$a_2 r^\alpha \leqslant \int G(u) \leqslant a_2 R^\alpha \qquad \forall u \in S$$

where $a_2 = 2\pi a_1$.

Now we are in position to evaluate $m = f(\bar{v})$. Let $\bar{v} = \lambda \bar{w}$, with $\bar{w} \in S$. Using (13) and (16) we get :

$$\lambda^{2-\alpha} \int \bar{w}\cdot L\bar{w} = \alpha \int G(\bar{w}) \geqslant \alpha a_2 r^\alpha$$

Recalling the definition of b we have :

(17) $$\lambda^{2-\alpha} \geqslant \frac{\alpha a_2 r^\alpha}{2b}$$

Using (12), (16) and (17), we find :

(18) $$m = f(\lambda\bar{w}) = \frac{2-\alpha}{2} \lambda^\alpha \int G(\bar{w}) \geqslant \frac{2-\alpha}{2} \lambda^\alpha a_2 r^\alpha \geqslant a_3 r^{2\alpha\delta} ,$$

with $$a_3 = \frac{2-\alpha}{2} \left(\frac{\alpha}{2b}\right)^{\alpha\delta} a_2^{2\delta}$$

Next, let $u = \psi(w)$ with $w \in S^n$. From $u = \lambda w$, it follows according to (13) and (16) :

(19) $$\lambda^{2-\alpha} \int w\cdot Lw = \alpha \int G(w) \leqslant \alpha a_2 R^\alpha$$

As remarked before, $b = \frac{1}{2} \int \bar{u} \cdot L u$ for some $\bar{u} \in S^n$ and hence $b = \frac{1}{2} \int w \cdot Lw$, $\forall\, w \in S^n$. Then, using again (12), (16) and (19) we find

$$f(\psi(w)) \leqslant a_3\, R^{2\alpha\delta}$$

with exactly the same a_3 as in (18). Hence :

$$\text{(20)} \qquad \tilde{m} := \max_{w \in S^n} f(\psi(w)) \leqslant a_3\, R^{2\alpha\delta}$$

Since $R^2 < 2r^2$, (20) implies

$$\tilde{m} < a_3\, 2^{\alpha\delta}\, r^{2\alpha\delta}$$

and by (18) :

$$\tilde{m} < 2^{\alpha\delta}\, m$$

From Lemma 9 we know that $2^{\alpha\delta} m = m^{\star}$ and therefore $\tilde{m} < m^{\star}$, as we wanted to show.

We end the paper with some remarks.

In their proof Ekeland-Lasry [10] make use of the same dual action principle, but substitute H with a suitable subquadratic Hamiltonian K. The corresponding functional is studied by a different method than that used here. The present proof shows the result is unaffected by the choice of the modified Hamiltonian, no matter if it is taken superquadratic of any degree of homogeneity.

It is also fairly transparent where the condition $R^2 < 2r^2$ plays its role : it ensures that $c_n < m^{\star}$ so that n critical levels of $f_{/M}$ are in a part where A is free. Or course, A is free not only in $\tilde{M}$ but in all $M \setminus M^{\star}$, and it is easy to find examples [20] in which f can have critical points, corresponding to solutions with minimal period

2π , at levels above $m^\star$. On the other hand, our estimates on $m^\star$ are sharp and cannot be improved : therefore it seems that, in order to obtain different multiplicity results, one should be able to give conditions which ensure, a priori, that the levels c_j have no intersection with $M^\star$.

REFERENCES

[1] A. Liapunov, Ann. Fac. Sci. Toulouse, 2 (1907), 203-474.

[2] A. Weinstein, Inv. Math., 20 (1973), 47-57.

[3] J. Moser, Comm. Pure Appl. Math., 29 (1976), 727-747.

[4] E. Hopf, Ber Math-Phis. Sachs. Acad. Wiss. Leipzig, 94 (1942), 1-22.

[5] J. Alexander and J. Yorke, Amer. J. Math., 100 (1978), 263-292.

[6] S.N. Chow and J. Mallet-Paret, J. Diff. Eq., 29 (1978), 66-85.

[7] E.R. Fadell and P.H. Rabinowitz, Inv. math. 45 (1978), 139-174.

[8] A. Weinstein, Annals of Math., 108 (1978), 507-518.

[9] P.H. Rabinowitz, Comm. Pure Appl. Math., 31 (1978), 157-184.

[10] I. Ekeland and J.M. Lasry, Annals of Math., 112 (1978), 157-184.

[11] F. Clarke and I. Ekeland, Comm. Pure Appl. Math., 33 (1980), 103-116.

[12] A. Ambrosetti and G. Mancini, Math. Annalen. 255 (1981), 405-421.

[13] G. Mancini, Periodic solutions of Hamiltonian systems having prescribed minimal period. Article in these Proceedings.

[14] A. Ambrosetti and P.H. Rabinowitz, Journal Funct. Anal., 14 (1973), 349-381.

[15] P.H. Rabinowitz, in Nonlinear Evolution Equations, M.G. Crandall Ed., Ac. Press (1978) 225-251.

[16] V. Benci and P.H. Rabinowitz, Inv. Math. 62 (1979), 336-352.

[17] F. Clarke, Periodic solutions to Hamilonian inclusions, to appear J. Diff. Eq.

[18] I. Ekeland, Jour. Diff. Eq., 34 (1979), 523-534.

[19] V. Benci, A geometrical index for the group S' and some applications to the study of periodic solutions of ordinary differential equations, to appear Comm. Pure Appl. Math.

[20] A. Ambrosetti and G. Mancini, On a theorem by Ekeland and Lasry concerning the number of periodic Hamiltonian trajectories. To appear Jour. Diff. Eq.

[21] Z. Nehair, Acta Math., 105 (1961), 141-175.

[22] C.V. Coffmann, J. Analyse Math., 22 (1969), 391-419.

[23] J.A. Hempel, Ind. Univ. Math., J., 20 (1971), 983-996.

[24] A. Ambrosetti, Atti Acc. Naz. Lincei, 52 (1972), 402-409.

[25] A. Ambrosetti, Rend. Sem. Mat. Univ. Padova, 49 (1973), 195-204.

[26] A. Ambrosetti and G. Mancini, in Recent Contributions to non-linear Partial Differential Equations, H. Berestycki and H. Brézis (Ed.), Pitman (1981), 24-36.

THE DIRECT METHOD IN THE STUDY OF PERIODIC SOLUTIONS

OF HAMILTONIAN SYSTEMS WITH PRESCRIBED PERIOD

V. BENCI

Instituto di Matematica Applicata - Università di Bari - Italy

Conference held in Rome
February 1981

Abstract.

The periodic solutions of Hamiltonian systems correspond to the critical points of the "action" functional which is indefinite in a "strong" sense. In this paper, we show the advantages and the difficulties of studying such indefinite functionals directly in an infinite dimensional function space. The outlines of some proofs are presented.

Note : Section 2 and 3 contain some results not presented in the conference held in Rome because of the more recent developments of the theory.

THE DIRECT METHOD IN THE STUDY OF PERIODIC SOLUTIONS OF HAMILTONIAN SYSTEMS WITH PRESCRIBED PERIOD

V. BENCI

I. - Introduction.

We consider the Hamiltonian system of 2n ordinary differential equations

$$(1.1) \qquad \dot{p} = -H_q(p,q) \qquad ; \qquad \dot{q} = H_p(p,q)$$

where $H \in C^2(\mathbb{R}^{2n},\mathbb{R})$; "$\cdot$" denotes $\frac{d}{dt}$ and $H_q = \frac{\partial H}{\partial q}$; $H_p = \frac{\partial H}{\partial p}$.

This system can be written more coincisely as

$$(1.2) \qquad -J\dot{z} = H_z(z)$$

where $z = (p,q)$ and $J = \begin{pmatrix} 0 & -Id \\ Id & 0 \end{pmatrix}$, Id being the identity in $\mathbb{R}^N$.

There are many types of questions, both local and global in the study of periodic solutions of (1.2) (cf. [R2] and its references and other conferences of this meeting).

In this conference I will be concerned with the existence of periodic solutions of (1.2) when the period $T = 2\pi\omega$ is precribed.

Making the change of variable $t \to \frac{1}{\omega} t$, (1.2) becomes

$$(1.3) \qquad -J\dot{z} = \omega H_z(z) \quad .$$

Of course, the 2π-periodic solutions of (1.3) correspond to the $2\pi\omega$-periodic solutions of (1.2).

The (1.3) are the Euler-Lagrange equations of the functional of the "action"

$$(1.4) \qquad f(z) = -\frac{1}{2}\int_0^{2\pi} (J\dot{z},z)\,dt - \omega\int_0^{2\pi} H(z)\,dt$$

This functional is difficult to study because it is indefinite in a "strong" sense. To be more precise, it is necessary to give some definitions. A functional f on a Banach space is called "definite" if it is bounded from below (or from above) ; "semidefinite" if there exists a weakly continuous function ϕ such that $f + \phi$ is definite. f is called "indefinite" if it is not semidefinite.

The spectrum of $z \to -J\dot{z}$ in $L^2(0,2\pi;\mathbb{R}^{2n})$ with periodic boundary conditions consits of infinitely many positive and negative eigenvalues. This fact makes the functional (1.4) indefinite in any reasonable function space in which we can study the problem. A different way to see the indefinite nature of the functional (1.4) is the following one. If (1.4) has a critical point it must be a saddle point whose negative and positive manifolds are both infinite dimensional.

Because of this indefiniteness the theory earlier developed for the study of critical points (cf. [Cl], [AR]) cannot be applied. Therefore some authors have studied this functional using indirect methods.

Rabinowitz, in his pioneering work, has applied a suitable finite dimensional approximation [Rl].

Amann and Zehnder ([A], [AZl] and [AZ2]) have applied a sort of Liapunov-Schmidt finite dimensional reduction method, which works if

$$(1.5) \qquad |H_{zz}(z)| \quad \text{is bounded for every} \quad z \in \mathbb{R}^{2n} .$$

If

$$(1.6) \qquad H \quad \text{is convex} \;,$$

It is possible to use a method introduced by Clarke [C] and Ekeland [E].

This method consists in studying the functional formally defined by

(1.7) $$f(v) = \psi^{\star}(v) - \frac{1}{2} \langle Kv,v\rangle$$

where K is the inverse of the operator $Lz = - Jz$ defined on $(\ker (L))$ and

$$\psi^{\star}(v) = \max_{u \in (\ker L)} \left\{\int_0^{2\pi} uv\, dt - \omega \int_0^{2\pi} H(u)\, dt\right.$$

If u is a critical point of (1.4), $w = Lu$ is a critical point of (1.7).

The functional (1.7) is easier to handle then (1.4) because it is semidefinite since $\psi^{\star}$ is bounded from below and K is compact in an appropriate function space.

I will not talk any more of this method since it has been exposed in other conferences of this meeting (cf. also their references).

I just want to observe that this method has been applied succes-sifully in the study of periodic solution of some nonlinear hyperbolic partial differential equations in cases in which other methods seem not to apply (cf. [BCN], [BF2] and their references).

If we want to avoid restrictions as (1.5) or (1.6), it is neces-sary to study the functional (1.4) directly.

Moreover the direct study of (1.4) permits us to develop a critical point theory for indefinite functionals, which, as we will see later, presents particular features if compared to the semidefinite functional theory.

2. - The direct method.

In this section I will try to show which are the main difficulties in studying (1.4) directly, and some of the available results (for other results cf. [BR], [B1], [FHR], [R3], [BCF1]).

We suppose that the function H satisfies the following growth condition

(2.1) there exist positive constants k_1, k_2, such that

$$|H_z(z)| \leqslant k_1 + k_2 |z|^{\alpha}$$

Because of this growth condition, the functional f is Fréchet differentiable in the Hilbert space $E = W^{1/2}(S^1, \mathbb{R}^{2n})$ $(S^1 = \mathbb{R}/_{2\pi\mathbb{Z}})$ of 2^n-ples of 2π-periodic functions, which possesses square integrable "derivative of order 1/2". Perhaps the simplest way to introduce this space is as follows. Let $z \in C^{\infty}(S^1, \mathbb{R}^{2n})$. Then z has a Fourier expansion $z = \sum_{-\infty}^{\infty} z_k e^{ikt}$. $W^{1/2}(S^1, \mathbb{R}^{2N})$ is the closure of the set of such functions with respect to the (Hilbert space) norm

$$\|z\| = \left| \sum_{k \in \mathbb{Z}} (1 + |k|) |z_k|^2 \right|^{1/2} .$$

On the Hilbert space E there is a unitary representation of the group

$$C_{\infty} = \{w \in C \mid |w| = 1\}$$

defined as follows :

$$\text{if } w = e^{is} \quad (s \in [0, 2\pi)) \quad , \quad (T_w z)(t) = z(t+s) .$$

The functional (1.4), defined on E (or, more generally, on any set of functions defined on S^1) is invariant with respect to this action, i.e.

$$f(z) = f(T_w z) \qquad \text{for every} \quad w \in C_\infty \ .$$

This fact permits us to use a topological invariant, called index, which is a function $i : \Gamma \to N \cup \{+\infty\}$ where Γ is the family of closed invariant subsets of E . For its definition and its properties we refer to the conference of Ambrosetti (cf. [B1] for further details ; cf. also [FR] and[FHR] for "homological" indices).

We need to add the following property :

$$\text{(2.2)} \qquad i(X) = +\infty \quad \text{if} \quad X \cap \text{Fix}\,(T_w) \neq \emptyset$$

where $\text{Fix}\,(T_w) = \{z \in E \mid T_w z = z \ \text{ for every } \ w \in C_\infty\}$.
(Ambrosetti did not mention this property since, in the case he has considered, $\text{Fix}\,(T_w) = 0$).

Unfortunately this index is not sufficient to apply the Lijustenik-Schnirelmann theory to indefinite functionals as (1.4).

In fact, if we put

$$c_k = \inf_{i(X) \geqslant k} \ \sup_{z \in X} \ f(z)$$

it happens that $c_k = -\infty$ for every $k \in \mathbb{N}$ (cf. the difference with the definite functional defined in the conference of Ambrosetti).

However it is possible to use this index to define a more sophisticated topological invariant called "pseudoindex".

Let $\mathcal{H}$ be a group of homeomorphisms on the Hilbert space E . We suppose that every $h \in \mathcal{H}$ is T_w-equivariant (i.e. $T_w \circ h = h \circ T_w$ for every $w \in C_\infty$). For every $X, Y \subset \Gamma$ we set

$$(2.3) \qquad i(X,Y) = \min_{h \in \mathcal{H}} i(h(X) \cap Y)$$

Some notation is now necessary. Given $c \in \mathbb{R}$, we set

$$K_c = \{z \in E \mid f(z) = c , f'(z) = 0\}$$

$$f^c = \{z \in E \mid f(z) \leqslant c\}$$

$$f_c = \{z \in E \mid f(z) \geqslant c\}$$

If $X \subset E$, we set $N_\delta(X) = \{z \in E \mid \text{dist}(z,X) < \delta\}$.

<u>2.4 Definition.</u> We say that the triple $\{f,\mathcal{H},(\alpha,\beta)\}$ $(\alpha,\beta \in \mathbb{R})$ satisfies the property (P) if, for every $c \in (\alpha,\beta)$ and for every $\delta > 0$, there exists $\varepsilon > 0$ and $\eta \in \mathcal{H}$ such that

$$\eta \, \overline{\left(f^{c+\varepsilon} - N_\delta(K_c)\right)} \subset f^{c-\varepsilon}$$

This definition allows us to state the following theorem.

<u>2.5 Theorem.</u> <u>Suppose that the triple</u> $\{f,\mathcal{H},(\alpha,\beta)\}$ <u>satisfies the property (P) and set</u>

$$(2.6) \qquad \bar{k} = i\left(f_{c_o} , f^{c_\infty}\right)$$

<u>where</u> $\alpha < c_o < c_\infty < \beta$.

<u>Moreover for</u> $k = 1,2,\ldots,\bar{k}$ <u>set</u>

$$\Gamma_k = \{X \in \Gamma \mid i\left(X,f_{c_o}\right) \geqslant k\}$$

<u>and</u>

$$(2.6') \qquad c_k = \inf_{X \in \Gamma_k} \sup_{z \in X} f(z)$$

<u>Then for every</u> $k = 1,\ldots,\bar{k}$, $c_k \in [c_o,c_\infty]$ <u>and it is a critical value</u>

of f. Moreover, if $c = c_k = c_{k+1} = \ldots = c_{k+r}$ $(k+r \leq \bar{k})$, then $i(K_c) \geq r+1$.

The proof of theorem 2.5 follows standard arguments, since the pseudoindex satisfies analogous property of Lijusternik-Schnirelmann category. Thus it will not be given here (cf. e.g. [B2] or [BCF1]).

The purpose of Theorem 2.5 is to show the difficulties in dealing with indefinite functionals. The main difficulty consists in determining the right group $\mathcal{H}$.

In fact, if this group is too large it may happen that $\bar{k}$ defined by (2.6) vanishes. This is the case if we try to apply Th. 2.5 to the functional (1.4) with $\mathcal{H}$ = {group of all equivariant homeomorphism on E} . On the other hand if $\mathcal{H}$ is too small, the triple $\{f, \mathcal{H}, (\alpha, \beta)\}$ does not satisfy the property (P). This happens if every $h \in \mathcal{H}$ has the form identity + compact. In any case the construction of the homeomorphism η as in Def. 2.4 and the evaluation of $\bar{k}$ given by (2.6) and (2.3), are delicate points (cf. [BR], [B2], [BCF1] and [FHR]).

In a recent work, [BCF1], the following result has been obtained.

2.7 Theorem. Let $\mathcal{H}$ be a real Hilbert space, on which a unitary representation T_w of the group C_∞ acts. Let $f \in C^1(E,\mathbb{R})$ be a functional on E satisfying the following assumptions :

f_1) $f(u) = \frac{1}{2}(Lu \mid u) - \psi(u)$, where $(\cdot|\cdot)$ is the linear product in E , L is bounded selfadjoint operator and $\psi \in C^1(E,\mathbb{R})$, $\psi(0) = 0$, is a functional whose Fréchet derivative is compact. We suppose that both L and ψ' are equivariant.

f_2) 0 does not belong to the essential spectrum of L.

f_3) Every sequence $\{u_n\} \subset E$, for which $f(u_n) \to c \in (0,+\infty)$ and $\|f'(u_n)\| \cdot \|u_n\| \to 0$, possesses a bounded subsequence.

f_4) <u>There are two closed subspaces</u> T_w<u>-invariant</u> $V, W \subset E$ <u>and</u> $R, \delta > 0$ <u>s.t.</u>

a) W <u>is L-invariant, i.e.</u> $LW = W$

b) $Fix(T_w) \subset V$ <u>or</u> $Fix(T_w) \subset W$

c) $f(u) < \delta$ <u>for</u> $u \in Fix(T_w)$

d) f <u>is bounded from above on</u> W

e) $f(u) \geq \delta$ <u>for</u> $u \in V$ <u>s.t.</u> $\|u\| = R$

f) $codim\,(V+W) < +\infty$, $dim\,(V \cap W) < +\infty$.

<u>Under the above assumptions there exists at least</u>

$$\frac{1}{2}\ (dim\,(V \cap W) - codim\,(V + W))$$

<u>independent critical points, with critical values greater or equal to</u> δ .

(Two critical points u_1 and u_2 are independent if $T_w u_1 \neq u_2$ for every $w \in C_\infty$).

<u>Outline of the proof.</u> Let $\mathcal{U}$ be the class of equivariant homeomorphisms $U : E \to E$ of the form

$$U = e^{\alpha(\cdot)L} .$$

where $\alpha : E \to \mathbb{R}$ is a G-invariant bounded functional.

We denote by $\mathcal{B}$ the class of continuous equivariant maps $b : E \to E$ such that for every bounded set $\Omega \subset E$, there exists a finite dimensional space $E_n \subset E$, spanned by a finite number of eigenvectors of L , such that

$$b(\Omega) \subset E_n$$

Finally we set

$$\mathcal{H} = \left\{ \begin{array}{l} h : H \to H \mid h \text{ is an equivariant homeomorphism such} \\ \text{that } h(0) = 0 \text{ , } h = U_1+b_1 \text{ , } h^{-1} = U_2+b_2 \text{ where} \\ U_1,U_2 \in \mathcal{U} \text{ and } b_1,b_2 \in \mathcal{B} \end{array} \right\}$$

It is not difficult to prove that $\mathcal{H}$ is a group.

By virtue of (f_1) (f_2) and (f_3) it is possible to prove that the triple $\{f,\mathcal{H},(0,+\infty)\}$ satisfies the property (P) by constructing a homeomorphism η as in Def. 2.4.

Moreover, using condition (f_4)(b), we get the following estimate of the pseudoindex :

$$\text{(2.8)} \qquad i(f^M,f_\delta) \geqslant \frac{1}{2}\,[\dim (V+W) - \operatorname{codim} (V+W)]$$

where $M = \sup f(W)$.

Both, the construction of η and the estimate (2.8) are very involved and they will not be given here.

Using theorem 2.5, it follows that the c_k's defined by (2.6') are critical values and bigger than δ . If they are all different from each other then the conclusion follows. If not, for some k , we have $i(K_c) \geqslant 2$ where $c = c_k$. Because of (f_4)(c), $K_c \cap \text{Fix}\,(T_w) = \emptyset$. Then, by virtue of a well-known property of the index (cf. [B1]) , K_c consists of infinitely many independent critical points. □

3 - Some applications of theorem 2.7.

Rabinowitz ([R1]) has proved that if $H(p,q)$ is "superquadratic" in both its variables p and q , i.e.

$$(3.1) \quad \begin{cases} \text{there exist constants } R > 0 \,,\ \mu > 2 \text{ such that} \\ (H_z(z),z) \geqslant \mu\, H(z) > 0 \qquad \text{for} \quad |z| > R \end{cases}$$

and if H satisfies some other technical assumptions, then 1.2 has a T-periodic solution for every $T > 0$ (later he actually proved that condition (3.1) by itself is sufficient to get this result [R3]).

After [R1], many other papers appeared dealing with (0.2) when H(z) is superquadratic (cf. [BR], [B2], [E], [CE], [AM], [R3] and their references).

Unfortunately, the above results on superquadratic Hamiltonian systems do not cover the classical mechanical problems. In fact, consider a mechanical system with constraints not depending on time, imbedded in a conservative field of forces. The Hamiltonian of such a system has the form

$$(3.2) \qquad H(r,q) = \sum_{ij=1}^{n} a_{ij}(q)\, p_i\, p_j + V(q)$$

where $\{a_{ij}(q)\}$ is a definite matrix for every $q \in \mathbb{R}^n$. The Hamiltonian (3.2) is quadratic in p , thus it does not satisfy (3.1).

If the a_{ij}'s do not depend on q , (1.1) can be transformed to a seconf order of n-equations of the form

$$(3.3) \qquad \ddot{x} = -\frac{\partial U(x)}{\partial x} \qquad x \in \mathbb{R}^n$$

Therefore in this case, the research of T-periodic solutions is reduced to the study of the critical points of the functional

$$(3.4) \qquad f(x) = \int_0^{2\pi} \left\{ \frac{1}{2} |\dot{x}|^2 - \omega U(x) \right\} dt \qquad x \in W^1(S^1, \mathbb{R}^{2n}) , \quad \omega = T/2\pi$$

The functional (3.4) is semidefinite, because it is bounded from below modulo the weakly continuous perturbation

$$x \to \omega \int_0^{2\pi} U(x)\, dt$$

In this case, applying the theory for semidefinite functional it is possible to prove that (3.3) has a nonconstant T-periodic solution for every $T > 0$ provided that T grows more than quadratically in the sense of (3.1) (cf. [BF1], [R2]).

The following theorem, (cf. [BCF2]) for Hamiltonians having the form (3.2), is a particular case of a more general theorem proved in [BCF1].

<u>3.5 Theorem.</u> <u>Consider the Hamiltonian system</u> (1.1) <u>with</u> H(p,q) <u>having the form</u> (3.2). <u>Suppose that</u> $V \in C^1(\mathbb{R}^n, R)$ <u>and</u> $a_{ij} \in C^1(\mathbb{R}^n, R)$ $(i,j = 1,\dots,n)$ <u>satisfy the following assumptions</u> :

i) <u>there are constants</u> $c, r > 0$ <u>and</u> $\alpha > 2$ <u>s.t. for</u> $|q| > r$

$i_1)$ $0 < \alpha V(q) \leq (V_q, q)_{\mathbb{R}^n}$

$i_2)$ $|V_q(q)| \leq c|q|^\alpha$

ii) <u>there exists</u> $\nu > 0$ <u>s.t.</u>

$$\sum_{ij} a_{ij}(q) p_i p_j \geq \nu |p|^2 \quad \underline{\text{for every}} \quad p, q \in \mathbb{R}^n$$

iii) <u>if</u> $p,q \in \mathbb{R}^n$

$$\sum_{ij} A_{ij}(q)\, p_i\, p_j \geqslant 0 ,$$

<u>where</u> $A_{ij}(q) = (\text{grad } a_{ij}(q) , q)_{\mathbb{R}^n}$

iv) $|\text{grad } a_{ij}(q)|$ <u>is bounded</u> $(i,j = 1,\dots,n)$.

<u>Under the above assumptions,</u> (1.1) <u>possesses a nonconstant T-periodic solution for each fixed period T</u> .

<u>Outline of the proof.</u>

Obviously the functional (1.4) satisfies assumption (f_1) with $E = W^{1/2}$, $\psi(z) = \int_0^{2\pi} H(z(t))\, dt$ (we shall take in the sequel $\omega = 1$), and L is the continuous selfadjoint operator in $W^{1/2}$ defined by the bilinear form

(3.6) $$(Lu|v) = \langle -J\dot{u}, v\rangle ,$$

where $u,v \in W^{1/2}$ and $\langle \cdot,\cdot \rangle$ denotes the canonical pairing between $W^{1/2}$ and its dual. The spectrum of L is given by the eigenvalues $\lambda_j = \frac{j}{1+|j|}$, $j \in \mathbb{Z}$, λ_j has multiplicity 2n and dim Ker L = 2n . So also assumption (f_2) is satisfied.

Now we have to verify that (f_3) is satisfied. Let $\{z_n\} \subset W^{1/2}$ s.t. $f(z_n) \to c \in (0,+\infty)$ and $\|f'(z_n)\| \cdot \|z_n\| \to 0$. The first step, which will be not given here, is to prove that, by virtue of (i_1), (ii) and (iii), the sequences

(3.7) $$\|p_n\|_{L^2} , \quad \|q_n\|_L$$

are bounded. Hence $\|z_n\|_{L^2 \times L^\alpha}$ is bounded .

The second step is to prove that $\|z_n\|_{W^{1/2}}$ is bounded.

By (i_2) and (iv) some computations show that $\| H_z(z_n)\|_{L^1}$ is bounded. Then, if $\eta > 0$, by the Sobolev imbedding theorems that the sequence

$$(3.8) \qquad \| H_z(z_n) \|_{W^{-1/2-\eta/2}}$$

is bounded. On the other hand we know that $-J\dot{z}_n - H_z(z_n) \to 0$ in $W^{-1/2}$. Then by (3.8) we deduce that the sequence

$$(3.9) \qquad - J\, \dot{z}_n$$

is bounded in $W^{-1/2-\eta/2}$. Let E^+ (resp. E^-) be the span of the eigenfunctions corresponding to the positive (resp. negative) eigenvalues and $E^o = \text{Ker } L$. Then $z = z^+ + z^- + z^o = \tilde{z} + z^o$ with $z^+ \in E^+$, $z^- \in E^-$, $z^o \in E^o$. It is easy to show that $\|z\|_{W^{+1/2-\eta/2}} \leqslant \text{const.}\ \|J\dot{z}\|_{W^{-1/2-\eta/2}}$. Then by (3.9) we deduce that $\|\tilde{z}\|_{W^{+1/2-\eta/2}}$ is bounded. Moreover since Ker L is finite-dimensional, we have also that $\|z_n\|_{W^{+1/2-\eta/2}}$ is bounded. Hence, as $\eta > 0$ is arbitrary, by Sobolev embedding theorems we deduce that for every $t > 1$

$$(3.10) \qquad \|z_n\|_{L^t \times L^t}$$

is bounded. Now, as $\langle f'(z_n), z_n\rangle \to 0$, it is easy to see that for $n \in \mathbb{N}$

$$(3.11) \qquad \|z_n^+\|^2_{W^{1/2}} \leqslant \text{const} \left(1 + \int_0^{2\pi} (H_z(z_n), z_n^+)_{\mathbb{R}^{2n}}\, dt\right)$$

Moreover by (i_2) and (iv) we deduce that for $n \in \mathbb{N}$

$$(3.12) \qquad \int_0^{2\pi} (H_z(z_n), z_n^+)_{\mathbb{R}^{2n}}\, dt \leqslant \text{const} \left(1 + \|z_n\|^t_{L^{2t}} \cdot \|z_n^+\|_{L^2}\right),$$

where $t = \max\{2,\alpha\}$. By (3.11) , (3.12) and (3.10) we conclude that $\|z_n^+\|_{W^{1/2}}$ is bounded. Analogously, we prove that $\|\bar{z}_n\|_{W^{1/2}}$ is bounded. Finally because Ker L is finite-dimensional, we have also that $\|z_n^o\|_{W^{1/2}}$ is bounded.

In order to verify the "geometrical" assumption (f_4), we need more technicality. Here we shall state the main steps of this proof only :

I step) Let $\delta > 0$ be such that

$$\delta \geqslant 2\pi \cdot \sup\{-H(z) \mid z \in \mathbb{R}^{2n}\} \tag{3.13}$$

In correspondence to δ , we can choose $R > 0$ and $j \in \mathbb{N}$ sufficiently large in order that $f(z) > \delta$ for every $z \in E^+$, $\|z\|_{W^{1/2}} = \mathbb{R}$, where E_k^+ is the span of the eigenfunctions corresponding to the eigenvalues λ_j , $j > k$. We set $V = E_k^+$.

II step) We set $W = (E_{k+n}^+)$, with $n \in \mathbb{N}$, $n \geqslant 1$. Using the superquadratic growth of $V(q)$ it can be proved that f is bounded from above on W . So the assumptions of the Theorem 2.7 are fulfilled. Now observe that by Theorem 2.7, the critical values which we find are greater or equal than δ . So by (3.13) we have that the corresponding critical points are not constants. Therefore we find $\frac{1}{2}$ (dim $(V \cap W)$ - codim $(V + W)$) = n non constant critical points.

Observe, moreover, that by the fact that n is arbitrary, we can deduce also that these critical points are infinite. So there exist infinitely many non constant T-periodic solutions of (1.2).

Now we will consider an other application of Th. 2.7 in the case in which H is asymptotically quadratic, i.e. there exists a linear operator $H_{zz}(\infty) : \mathbb{R}^{2n} \to \mathbb{R}^{2n}$

$$H_z(z) = H_{zz}(\infty)\, z + O(|z|) \qquad \text{for} \quad |z| \to +\infty . \tag{3.14}$$

The aim is to give a lower bound to the number of $2\pi\omega$-periodic solution by comparing of the operators $H_{zz}(0)$ and $H_{zz}(\infty)$.

As in [B2] , we define an even integer number $\theta(\omega H_{zz}(0), \omega H_{zz}(\infty))$ which will provide such bound.

Given the Hermitian operators $A,B : C^{2n} \to C^{2n}$ we set

$$N(A) = \{\text{number of negative eigenvalues of } A\}$$

$$\overline{N}(A) = \{\text{number of nonpositive eigenvalue of } A\}$$

and

$$\theta(A,B) = \sum_{k \in \mathbb{Z}} [N(ikJ + A) - \overline{N}(ikj + B)] .$$

Observe that $\theta(A,B)$ is a finite number. In fact for k big enough $N(ikJ + A) = \overline{N}(ikj + B) = n$.

Let $\sigma(A)$ denote the spectrum of the Hermitian matrix A .
If

(3.15) $$\sigma(i\,\omega\, J\, H_{zz}(\infty)) \cap \mathbb{Z} = \emptyset$$

and

(3.16) $$\sigma(i\,\omega\, J\, H_{zz}(0)) \cap \mathbb{Z} = \emptyset$$

then $\theta(\omega\, H_{zz}(\infty) , \omega\, H_{zz}(0)) = -\theta(\omega\, H_{zz}(0)\ \omega\, H_{zz}(\infty))$.

<u>3.18 Theorem.</u> <u>Suppose that</u> H <u>satisfies</u> (3.14) <u>and</u> (3.15).

<u>Moreover suppose that</u>

(3.18) $H_{zz}(0)$ <u>and</u> $H_{zz}(\infty)$ <u>are positive definite</u>

(3.19) $H_z(z) \neq 0$ <u>for every</u> $z \in \mathbb{R}^{2n}$

<u>Then</u> (3.2) <u>has at least</u> $\frac{1}{2}\max\{\theta(\omega H_{zz}(\infty),\omega H_{zz}(0)) , \theta(\omega H_{zz}(0),\omega H_{zz}(\infty))\}$ <u>non constant</u> $2\pi\omega$-<u>periodic solutions</u>.

Th. 3.18 is a consequence of a more general theorem proved in [BCF1] (cf. [B2] for a variant of this theorem).

Amann and Zehnder [AZ2], using the finite reduction method have obtained a similar result, using, instead of (3.18) the stronger assumption of uniform convexity of H .

REFERENCES

[A] H. AMANN , Saddle points and multiple solutions of differential equations, Math., Z. 169, (1979), 127-166.

[AZ1] H. AMANN - E. ZEHNDER , Nontrivial solutions for a class of nonresonance problems and applications to nonlinear differential equations, Ann. Sc. Norm. Sup. Pisa, in press.

[AZ2] H. AMANN - E. ZEHNDER , Periodic solutions of asymptotically linear Hamiltonian systems, Preprint.

[AM] A.AMBROSETTI - G. MANCINI , Solutions of minimal period for a class of convex Hamiltonian systems, Preprint.

[AR] A. AMBROSETTI - P.H. RABINOWITZ , Dual variational methods in critical point theory and applications, J. Funct. Anal. 14 , (1973), 349-381.

[B1] V. BENCI , A geometrical Index for the group S^1 and some applications to the study of periodic solutions of ordinary differential equations, comm. Pure Appl. Math., 34 (1981), 393-432.

[B2] V. BENCI , On the critical point theory for indefinite functionals in the presence of symmetries, to appear in Trans. Amer. Math. Soc.

[BCF1] V. BENCI - A. CAPOZZI - D. FORTUNADO , Periodic solutions of Hamiltonian systems with a prescribed period, Preprint.

[BCF2] V. BENCI - A. CAPOZZI - D. FORTUNADO , Periodic solutions for a class of Hamiltonian systems, to appear.

[BF1] V. BENCI - D. FORTUNADO , Un teorema di molteplicità per un'equazione ellittica non lineare su varietà simmetriche, Proceedings of the Symposium "Metodi asintotici e topologici in problem diff. non lineari", L'Aquila (1981).

[BF2] V. BENCI - D. FORTUNADO , The dual method in critical point theory. Multiplicity results for indefinite functionals, to appear in Ann. Mat. Pura e Applicata.

[BR] V. BENCI - P.H. RABINOWITZ , Critical point theorems for indefinite functionals, Inv. Math., 52 (1979), 336-352.

[BCN] H. BREZIS - J.M. CORON - L. NIRENBERG , Free vibrations for a nonlinear wave equation and a theorem of P. Rabinowitz, Preprint.

[Cl] CLARK B.C. , A variant of Ljiusternik Schnirelmann theory, to appear in J. Diff. Eq.

[CE] F.H. CLARKE - I. EKELAND , Hamiltonian trajectories having prescribed minimal period, comm. Pure Appl. Math., 33, (1980), 103-116.

[E] I. EKELAND , Periodic solutions of Hamiltonian equations and a theorem of P. Rabinowitz, J; Diff. Eq., 34, (1979), 523-534.

[FHR] E.R. FADELL - S. HUSSEINI - P.H. RABINOWITZ , Borsuk-Ulam theorems for arbitrary S^1 actions and applications, Math. Research Center Technical Summary Report, University of Wisconsin-Madison, 1981.

[FR] E.R. FADELL - P.H. RABINOWITZ , Generalized cohomological index theories for Lie "group actions with an application to bifurcation questions for Hamiltonian systems, Inv. Math., 45, (1978), 139-174.

[R1] P.H. RABINOWITZ , Periodic solutions of Hamiltonian systems, Comm. Pure Appl. Math., 31 , (1978), 157-184.

[R2] P.H. RABINOWITZ , Periodic solutions of Hamiltonian systems : a survey, Math. Research Center Technical Summary Report, University of Wisconsin-Madison.

[R3] P.H. RABINOWITZ , Periodic solutions of large norm of Hamiltonian systems, Math. Resaerch Center Technical Summary Report, University of Wisconsin-Madison (1981).

PERIODIC SOLUTIONS OF HAMILTONIAN SYSTEMS HAVING PRESCRIBED MINIMAL PERIOD

Giovanni MANCINI

Universita di Trieste

RESUME

En partant des travaux de F. Clarke et I. Ekeland, on écrit une version en dimension finie de leur principe de dualité (cela est obtenu par une méthode de discrétisation).

De ce résultat on déduit un algorithme qui permet de trouver des solutions périodiques approchées à des équations différentielles non-linéaires.

On illustre cette méthode par les résultats obtenus pour l'équation :

$$\frac{d^2x}{dt^2} - x^3 + \alpha\,(\cos t + \sin t) = 0 \qquad x(\cdot) \in \mathbb{R}$$

ABSTRACT

Using the works of F. Clarke and I. Ekeland, one shows a similar dual principle in the finite dimensionnal case, by discretisation.

From this result, we infer an algorithm (which has been computed), to find periodic solutions to non-linear differential equations.

Finally, we illustrate this method by some numerical results concerning the equation :

$$\frac{d^2x}{dt^2} - x^3 + \alpha\,(\cos t + \sin t) = 0 \qquad x(\cdot) \in \mathbb{R}$$

KEY-WORDS

NUMERICAL ANALYSIS, NON LINEAR DIFFERENTIAL EQUATION, NON CONVEX OPTIMISATION, FORCED OSCILLATION.

PERIODIC SOLUTIONS OF HAMILTONIAN SYSTEMS HAVING PRESCRIBED MINIMAL PERIOD

Giovanni Mancini
Universita di Trieste

0. Introduction

Much progress has been done in recent years in the study of periodic solutions of Hamiltonian systems and many new ideas and methods of nonlinear functional analysis have been developed in this connection. Periodic orbits near an equilibrium as well as existence in the large of closed trajectories are the main topics where variational and topological methods have produced the deepest results. A strong motivation should be found in important papers by Weinstein [1] and Moser [2]. Improving a classical result of Lyapunov [3], they have shown that the autonomous hamiltonian system with N degrees of freedom

$$\text{(H)} \qquad \dot{p} = -\frac{\partial H}{\partial q}(p,q) \quad , \quad \dot{q} = \frac{\partial H}{\partial p}(p,q) \quad , \quad p,q \in C^1(\mathbb{R},\mathbb{R}^N)$$

has at least N distinct periodic solutions on each energy surface $H(\zeta) = \varepsilon^2$, ε small, provided $H \in C^2(\mathbb{R}^{2N},\mathbb{R})$ has a stationary point at $\zeta = 0$ and $H''(0)$ is positive definite. Again in the framework of bifurcation analysis, but choosing as a parameter the unknown period, Fadell and Rabinowitz [4] have given a more precise multiplicity result.

Both results quoted above are local in nature, and very little is known about similar multiplicity results in the large (see Ekeland-Lasry [5] for a partial global extension of the Weinstein-Moser Theorem, and Ambrosetti and Mancini [6] for an alternative proof ; see also Amann and Zehnder [7] for asymptotically linear hamiltonian systems). Nevertheless, restricting attention to the existence of a one parameter family of periodic solutions, rather satisfactory results have been obtained. In [8] Rabinowitz has proved the existence of closed trajectories for (H) lying on an energy surface $S = H^{-1}(1)$, provided $\nabla H(\zeta) \neq 0 \;\forall\; \zeta \in S$ and S is radially diffeomorphic to S^{2N-1}. In a standard fashion, he first considers the nonlinear eigenvalue problem

$$(H)_\lambda \qquad \dot{p} = -\lambda \frac{\partial H}{\partial q} \quad , \quad \dot{q} = \lambda \frac{\partial H}{\partial p}$$

where now p and q are 2π periodic functions and λ is a real parameter. After a change of the time variable, solutions to $(H)_\lambda$ give $2\pi\lambda$ periodic solutions of (H). Then he looks for critical points of the action integral $\int_0^{2\pi} \langle p,\dot{q}\rangle_{\mathbb{R}^N}\, dt$ subject to the constraint $(1/2\pi)\int_0^{2\pi} H(z(t))\, dt = 1$. In this connection, he deeply exploits the symmetry properties of $(H)_\lambda$ to get appropriate critical points of minimax type.

In a subsequent paper [9] Rabinowitz gives an alternative proof of the above result. It is derived, by means of a simple device, from the existence of periodic solutions of given period T , for a suitably modified hamiltonian system. Here, we will just deal with families of periodic solutions of (H) parametrized by the unknown period T . In contrast with the fixed energy case, one is led to look for free critical points of the functional

$$\int_0^{2\pi} \langle p,\dot{q}\rangle_{\mathbb{R}^N}\, dt - \int_0^{2\pi} H(z(t))dt \quad , \quad z = (p,q)$$

where now $\lambda = T/2\pi$ is a given real number. Assuming $\nabla H^1(0) = 0$, $(p,q) = (0,0)$ is a critical point and one looks for nontrivial ones. To this extent, very deep minimax arguments have been introduced by Rabinowitz [8] and further developed by Benci and Rabinowitz [10]. It should be remarked that no index theory, exploiting symmetries, is needed here.

However, it should be pointed out that such minimax arguments seem not to yield any information about the primitive period of the corresponding solution. A first result in this direction is given, using a different approach, by Clarke and Ekeland [11]; they prove that for every $T > 0$ (H) has a periodic solution having T as minimal period, provided H is convex and behaves "subquadratically" both at zero and at infinity. In section 1 below we give some results concerning the superquadratic case. Theorem 1.4 states the existence of periodic solutions having arbitrarily long minimal period. As a corollary we get, in a rather special geometric situation, solutions of any prescribed minimal period. We also present a result obtained jointly with Ambrosetti [12] concerning the existence of periodic solutions of small amplitude and long (minimal) period : a local result which is a consequence of a more complicated but global one (see theorem 1.7 below). In section 2 we describe a variational principle which allows us to find free critical points of saddle type as constrained minima : this is the key step in the proof of Theorem 1.7.

We wish finally to mention that, by means of continuation arguments, Alexander and Yorke [13] (see also recent work of Yorke and Mallet-Paret [14]) have proved the existence of global branches of periodic solutions of (H) bifurcating from stationary points. A much weaker results, for second order hamiltonian systems, was given earlier by Berger (see [15]).

In the last section of this exposition, we will consider a situation similar to the one in [15]. Assuming strong positivity on the Hessian matrix of the potential energy, and suitable growth conditions, we prove the existence of a continuous branch of periodic solutions, which are "increasing in half period". Along the branch the amplitude of the trajectories goes to infinity while the (minimal) period goes to zero.

1. Solutions of Prescribed Minimal Period : A Variational Approach.

Denote by $|\cdot|$, $<\cdot,\cdot>$ the norm and the euclidean scalar product, respectively, in $\mathbb{R}^{2N}$.

In [9] Rabinowitz has proved the following.

Theorem 1.1 (Rabinowitz [9]). Suppose $H \in C^1(\mathbb{R}^{2N},\mathbb{R})$ satisfies

(1.1) $$\exists r > 0 \ , \ \beta > 2 : \ 0 < \beta H(\xi) \leqslant <\xi,\nabla H(\xi)> \quad \forall |\xi| > r \ .$$

Then, $\forall$ T , R > 0 , (H) has a T-periodic solution z satisfying $\|z\|_{L^\infty} > R$.

Remark 1.2. As pointed out by Rabinowitz (see for example [16]), T might not be, in general, the real period of the solution given by Theorem 1.1. In fact, under the mere assumption (1.1), an upper bound, say T_o , for the minimal period of any solution of (H) could exist. This occurs, for example, if $H(\xi) = \phi(|\xi|^2)$, $\phi \in C^\infty$, $\phi'(z) > 0$, $\forall z > 0$ and $\phi'(0) = 1$, $\phi'(+\infty) = \infty$. The energy surfaces are spheres and the solutions of (H) (at the energy level $h \equiv H(z(t))$) are

$$\psi(h^{1/2})\left(\xi \cos\gamma(h)t + \eta \sin\gamma(h)t \ , \ \xi \sin\gamma(h)t + \right.$$
$$\left. - \eta \cos\gamma(h)t\right)$$

where $\xi,\eta \in \mathbb{R}^N$, $|\xi|^2 = |\eta|^2 = 1$, $\psi = \phi^{-1}$, $\gamma(h) = 2/\psi'(h)$. The corresponding periods, $\pi\,\psi'(h)$, are bounded from above. Thus, a periodic solution of "long" period T given by Theorem 1.1, could be nothing else than a solution of minimal period $T/k \leqslant T_o$. Hence Theorem 1.1 is, in some sense, a local result : it states the existence of solutions of (possibly) small period at high energies. We will give below, following Ekeland [17] a simple, and local, proof of it, under an additional convexity hypothesis on H.

We wish to remark that, while there are several other results concerning the existence of solutions for (H) of arbitrarily given period, much less is known about the existence of solutions of prescribed minimal period. For the sake of simplicity, we begin considering this problem in the very simple situation of a second order hamiltonian system

$$(1.2) \qquad \ddot{x} + \nabla U(x) = 0 \quad , \quad x \in C^2(\mathbb{R},\mathbb{R}^N)$$

where $U \in C^1(\mathbb{R}^N,\mathbb{R})$, $U(0) = 0$, $\nabla U(0) = 0$, is assumed to be even. In this case, we are led to look for odd periodic solutions. It amounts to finding the critical points of the functional

$$(1.3) \qquad I_\lambda(x) = \frac{1}{2}\int_0^\pi |\dot{x}|^2 - \lambda^2 \int_0^\pi U(x) \quad , \quad x \in H_0^1([0,\pi],\mathbb{R}^N)$$

where λ is a given real parameter. If $\bar{x}$ is a critical point of I_λ in H_0^1, it is a solution of the boundary value problem

$$(1.2)_\lambda \qquad \ddot{x} + \lambda^2 \nabla U(x) = 0 \quad , \quad x(0) = x(\pi) = 0$$

Setting $x(t) = \bar{x}(t/\lambda)$, we can extend it to $[-\pi,0]$ by oddness and then to all of $\mathbb{R}$ by periodicity, to get a $2\pi\lambda$ periodic solution of (1.2). Now, if U is subquadratic, i.e., $U(\xi)/|\xi|^2 \underset{\xi \to 0}{\to} \infty$, $U(\xi)|\xi|^{-2} \underset{|\xi| \to \infty}{\to} 0$, then it is easy to see that I_λ has a negative minimum, which is attained at some $\bar{x} \neq 0$. Finally, if x is its extension to all of $\mathbb{R}$ by oddness and periodicity, $\tilde{x}$ shall have minimal period 2π , otherwise $\tilde{x}_k(t) = \tilde{x}(t/k)$, $t \in [0,2\pi]$, should belong to H_0^1 for some integer $k > 1$; but

$$I_\lambda(\tilde{x}_k) = \frac{1}{2k^2}\int_0^{2\pi} |\dot{x}|^2 - \lambda^2 \int_0^\pi U(\bar{x}) < I_\lambda(\bar{x}) ,$$

a contradiction.

On the other hand, if U is superquadratic : $U(\xi)/|\xi|^2 \underset{|\xi| \to \infty}{\to} \infty$ and $U(\xi)/|\xi|^2 \underset{\xi \to 0}{\to} 0$, then I_λ is no longer bounded from below (as well as from above) but it has a strict local minimum at $x = 0$, and one should look for critical points of saddle type (using, for example, the Ambrosetti-Rabinowitz "mountain pass" Lemma [18]). Unfortunately, a direct argument as above does not apply here to prove the desired minimality property of the period. However, making some additional geometric assumptions on U, it is possible to find a free critical point of I_λ , minimizing it on the constraint $M = \{x \in H_0^1 : x \neq 0 , (\nabla I_\lambda(x),x)_{H_0^1} = 0\}$. Just as in the case of a free minimum, we get a solution of our problem. A further difficulty arises if one drops the evenness assumption on U. Exploiting the invariance of (1.2) under reflection in time, one can look for even periodic solutions of (1.2), i.e. for critical points of $I_\lambda(x)$, but with x belonging to the wider space H^1 . However, in H^1 , even in the subquadratic case, I_λ is no longer bounded from below, nor has a local minimum at zero. Nevertheless, arguments similar to the ones sketched above, used in connection with a "dual" formulation of the variational problem, apply to get solutions of prescribed minimal period, even in the more general framework of first order systems.

After these preliminary heuristics, let us now list the main results.

Theorem 1.3 (Clarke-Ekeland [11]). Let $H \in C^1(\mathbb{R}^{2N},\mathbb{R})$ be convex, and $H(0) = 0$, $\nabla H(0) = 0$. Assume

$$(1.4) \qquad \frac{H(\xi)}{|\xi|^2} \underset{|\xi| \to \infty}{\to} 0 \quad , \qquad \frac{H(\xi)}{|\xi|^2} \underset{|\xi| \to 0}{\to} \infty$$

Then, $\forall\, T > 0$, (H) has a periodic solution having T as minimal period.

As for superquadratic hamiltonians, we state a theorem which, despite the restrcitive growth assumptions, is in fact a first step to get more general results, when combined with approximation arguments.

Theorem 1.4. Let $H \in C^1(\mathbb{R}^{2N},\mathbb{R})$ be strictly convex. Assume there exists $\beta > 2$ such that :

(1.5) $\quad m|\xi|^\beta \leqslant \beta H(\xi) \leqslant M\,|\xi|^\beta \qquad \forall\, \xi \in \mathbb{R}^{2N}$,
for some constants $0 < m < M$

(1.6) $\quad 0 < \beta H(\xi) \leqslant \langle \nabla H(\xi), \xi \rangle \qquad \forall\, \xi \in \mathbb{R}^{2N}$, $\xi \neq 0$.

Then, for every $T > 0$, (H) has a T-periodic solution x_T, whose energy $h_T \equiv H(z_T(t))$ satisfies the inequalities

(1.7) $\quad c\,T^{\beta/2-\beta} \leqslant h_T \leqslant C\,T^{\beta/2-\beta}$

where c , C are constants only depending on m, M and β . Furthermore, if $K > \frac{\beta}{\beta-1}\,(M/m)^{1/\beta-1}$, then z_T cannot be Tk^{-1} periodic .

Corollary 1.5. Under the assumptions of Theorem 1.4, if $M < 2^{\beta-1}\left(1 - \frac{1}{\beta}\right)^{\beta-1} m$, then (H) has solutions of any prescribed minimal period.

A more general, but local, result concerning periodic solutions of any long minimal period for superquadratic hamiltonians, is the following.

Theorem 1.6 (Ambrosetti-Mancini [12]). Let $H(\xi) = \overline{H}(\xi) + R(\xi)$, where $R, \overline{H} \in C^2(\mathbb{R}^{2N}, \mathbb{R})$ satisfy

(i) $\overline{H}$ is strictly convex and there exists some $\beta > 2$ such that $\overline{H}(t\xi) = t^\beta \overline{H}(\xi)$, $\forall\, t \geqslant 0$, $\xi \in \mathbb{R}^{2N}$;

(ii) $R(0) = 0$, $\nabla R(0) = 0$, $\|R''(\xi)\| = 0\left(|\xi|^{\beta-2}\right)$ as $\xi \to 0$.

Then $T_o > 0$ such that for every $T \geqslant T_o$ (H) has a solution having T as minimal period. The amplitude of such solutions tends to zero as T tends to infinity.

Now, we will sketch the "dual variational principle", which we will use to study (H). For the sake of simplicity, we assume in what follows that (1.5) holds for some $\beta > 1$; in the subquadratic case, where obviously β should be less than 2, (1.5) will replace (1.4).

Let me now introduce some notation. Setting $\mathcal{I} = \begin{pmatrix} 0 & -I \\ I & 0 \end{pmatrix}$, where I denotes the identity N x N matrix, equation (H_λ) can be rewritten

$$(H)_\lambda \qquad \dot{z} = \lambda \mathcal{I} \nabla H(z) \quad , \quad z = (p,q)$$

Let A be the operator defined in $\text{dom } A = \{z \in H^{1,\alpha}(0,2\pi:\mathbb{R}^{2N}) : z(0) = z(2\pi)\}$, $\alpha^{-1} + \beta^{-1} = 1$, by the equation $Az = -\mathcal{I}\dot{z}$. Notice that $\text{Ker } A \equiv \mathbb{R}^{2N}$. Set $E = \left\{U \in L^\alpha(0,2\pi;\mathbb{R}^{2N}) : \int_0^{2\pi} u = 0\right\}$ and denote by $\|\cdot\|$ the norm in $L^\alpha(0,2\pi:\mathbb{R}^{2N})$. A , as an operator from $\text{dom } A \cap E$ into E , has a continuous inverse, which we call L. Note that L is a compact operator from E into $L^\alpha(0,2\pi;\mathbb{R}^{2N})$.

Now, in order to state the "dual" formulation of the variational problem related to (H), let me remark that, by the assumption on H, it follows that ∇H is a global homeomorphism, whose inverse is the gradient of the Legendre transform G of H, given by the Fenchel formula

$$G(\eta) = \sup_\xi \{\langle\eta,\xi\rangle - H(\xi)\}$$

Notice that $G(0) = 0$, $\nabla G(0) = 0$, while (see [17] for the details) (1.5)-(1.6) imply

$$(1.8) \quad \begin{cases} m_1|\eta|^\alpha \leqslant G(\eta) \leqslant m_2|\eta|^\alpha & \forall\, \eta \in \mathbb{R}^{2N} , \\ \quad \text{where } \alpha^{-1}+\beta^{-1} = 1 , \ m_1 = \alpha^{-1}M^{-\alpha/\beta} , \ m_2 = \alpha^{-1}m^{-\alpha/\beta} ; \\ \alpha\, G(\eta) \geqslant \langle\eta,\nabla G(\eta)\rangle & \forall\, \eta \in \mathbb{R}^{2N} , \text{ or, equivalently,} \\ \quad G(t\eta) \leqslant t^\alpha G(\eta) & \forall\, t \geqslant 1 ,\ \eta \in \mathbb{R}^{2N} \\ \exists c_1 > 0 : |\nabla G(\eta)| \leqslant c_1|\eta|^{\alpha-1} & \forall\, \eta \in \mathbb{R}^{2N} \end{cases}$$

Also notice that $H(\xi) + G(\eta) = \langle\xi,\nabla H(\xi)\rangle = \langle\eta,\nabla G(\eta)\rangle$ if $\eta = \nabla H(\xi)$. From (1.8) it follows that $u \to \int_0^{2\pi} G(u)$ is a well defined C^1 functional on E.

The "dual variational principle", first introduced by Clarke [19], and subsequently used in several papers ([5], [11], [12], [20], to mention only a few) reads as follows :

Dual variational principle. Let $\lambda \in \mathbb{R}^+$ be given. If $u \in E$ is a critical point of

$$(1.9) \qquad J_\lambda(u) = -\frac{\lambda}{2}\int_0^{2\pi} \langle u, Lu\rangle + \int_0^{2\pi} G(u) \quad , \quad u \in E$$

then $z = \nabla G(u)$ is a 2π-periodic solution of $(H)_\lambda$.

This easily follows from the fact that for some $\xi \in \mathbb{R}^{2N}$ it results $-\lambda Lu + \nabla G(u) = \xi$ and so $\lambda \nabla H(z) = \lambda u = A(\lambda Lu + \xi) = Az$.

Sketch of the proof of Theorem 1.3 : Here $\beta \in (1,2)$. This implies that $\int G(u)$ dominates the quadratic part for large u and so J_λ (given by (1.9)) is bounded from below. Using compactness and convexity we see that J_λ achieves its negative minimum at some u_m. Since necessarily $\int_0^{2\pi} \langle u_m, Lu_m\rangle > 0$, the same device used for I_λ (given by (1.3)), implies u_m has minimal period 2π : $\nabla G\left(u\left(\frac{t}{\lambda}\right)\right)$ is a solution of (H) with minimal period $T = 2\pi\lambda$.

We now consider, following [17], the superquadratic case, namely $\beta > 2$. In this case J_λ is unbounded from below (as well as from above) but $\int_0^{2\pi} G(u)$ is dominating near $u = 0$; and so J_λ has a strict local minimum at zero. Since (1.6) implies J_λ satisfies a Palais-Smale condition, the Ambrosetti-Rabinowitz theorem applies to get a nontrivial critical point of minimax type for J_λ . These arguments, which follow [17], prove the existence part of Theorem 1.4.

Proof of Theorem 1.4 completed : For a given $T > 0$, we have already seen that (H) has a T-periodic solution corresponding to a critical point v_λ of saddle type of J_λ , where $2\lambda\pi = T$. The estimates (1.7) rest upon the characterization of $J_\lambda(v_\lambda)$ as a minimax :

$$J_\lambda(v_\lambda) = \inf_{\sigma \in \Sigma} \max_{v \in \sigma} J_\lambda(v)$$

where Σ is the family of continuous paths in E joining zero to a given point $e \neq 0$ satisfying $J_\lambda(e) \leqslant 0$. We first show that :

$$(1.10) \qquad dM^{\alpha/(2-\beta)}\, T^{\beta/(2-\beta)} \leqslant J_\lambda(v_\lambda) \leqslant d\, m^{\alpha/(2-\beta)}\, T^{\beta/(2-\beta)}$$

where $d > 0$ is some pure constant. We derive (1.10) by comparing J_λ with suitable simpler functionals. Set $G_i(\eta) = m_i|\eta|^\alpha$ (m_i being the same constants appearing in (1.8)). Thus $G_1(\eta) \leqslant G(\eta) \leqslant G_2(\eta)$, $\forall\, \eta \in \mathbb{R}^{2N}$.

Denote

$$J_\lambda^{(i)}(v) = -\frac{\lambda}{2}\int_0^{2\pi} \langle v,Lv\rangle + \int_0^{2\pi} G_i(v) \quad , \quad v \in E \ , \ i = 1,2$$

$$\Lambda^{(i)}(v) = -\lambda\int_0^{2\pi} \langle v,Lv\rangle + \int_0^{2\pi} \langle \nabla G_i(v),v\rangle$$

$$= -\lambda\int_0^{2\pi} \langle v,Lv\rangle + \alpha\int_0^{2\pi} G_1(v)$$

$$M_\lambda^{(i)} = \{v \in E : v \neq 0 \ , \ \Lambda_\lambda^{(i)}(v) = 0\}$$

It is easy to see (compare with section below for a more general situation and [6] for the details) that $J_\lambda^{(i)}(v)$, $v \in M_\lambda^{(i)}$, achieves its minimum at some $v_i \in M_\lambda^{(i)}$. We have

$$(1.11) \qquad J_\lambda^{(1)}(v_1) \leqslant J_\lambda(v_\lambda) \leqslant J_\lambda^{(2)}(v_2)$$

In fact, obviously, minimax $J_\lambda^{(1)} \leqslant$ minimax $J_\lambda \leqslant$ minimax $J_\lambda^{(2)}$. But, if $J_\lambda^{(2)}(e) \leqslant 0$, as we can suppose, then minimax $J_\lambda^{(i)} = J_\lambda^{(i)}(v_i)$, because any path in Σ meets $M_\lambda^{(i)}$ and $J_\lambda^{(i)}$ has a global maximum at v_i along the ray $\{tv_i : t \geqslant 0\}$ (notice that one can join $\bar{t}v_i$, where $J_\lambda^{(i)}(\bar{t}\, v_i) \leqslant 0$, and e with a path lying in the region where $J_\lambda^{(i)}$ is nonpositive).

Now, an easy computation (see [12]) shows that a pure constant ℓ exists such that $\ell(m_1/T)^{\beta/(\beta-2)} \leqslant J_\lambda^{(1)} v_1)$ and $J_\lambda^{(2)}(v_2) \leqslant \ell(m_2/T)^{\beta/(\beta-2)}$. Hence, using (1.11), we get (1.10).

To relate $J_\lambda(v_\lambda)$ to $h_T \equiv H(\nabla G(v_\lambda(t)))$, we recall that $H(z(t)) + G(v_\lambda(t)) = \langle \nabla H(z(t)), z(t)\rangle = \langle \nabla G(v_\lambda(t)), v_\lambda(t)\rangle \quad \forall\, t$, where $z(t) = \nabla G(v_\lambda(t))$. Furthermore, since $0 = \langle J'(v_\lambda, v_\lambda)\rangle_{E',E} = -\lambda \int \langle v_\lambda, Lv_\lambda\rangle + \int \langle \nabla G(v_\lambda, v_\lambda)\rangle$, we have $J_\lambda(v_\lambda) = \int G(v_\lambda) - \frac{1}{2}\langle \nabla G(v_\lambda, v_\lambda)\rangle$. Hence $H(z(t)) = -2 J_\lambda(v_\lambda) + \int_0^{2\pi} G(v_\lambda)$. But $J_\lambda(v_\lambda) \geqslant \left(1-\frac{\alpha}{2}\right)\int G(v_\lambda)$ by (1.8), while (see [17]) $H(\xi) \geqslant b\, \langle \xi, \nabla H(\xi)\rangle$ for some $b > 0$ only depending on m, M and β. Henceforth, since $\int \langle z, \nabla H(z)\rangle = \int \langle v_\lambda, \nabla G(v_\lambda)\rangle \geqslant \int G(v_\lambda) \geqslant J_\lambda(v_\lambda)$, we get

$$b\, J_\lambda(v_\lambda) \leqslant \int_0^{2\pi} H(z(t)) \leqslant 2\,\frac{\alpha-1}{2-\alpha}\, J_\lambda(v_\lambda),$$

and (1.7) follows in view of (1.10).

Finally, we prove the last statement in Theorem 1.4. Assuming v is $2\pi k^{-1}$ periodic for some $k \in \mathbb{N}$, we first prove

$$\text{(1.12)} \qquad \min_{\Sigma} \max_{\sigma} J_{\lambda/k} \leqslant \alpha^{\alpha/(2-\alpha)} J_\lambda(v_\lambda)$$

Notice that $\tilde{v}(t) = v_\lambda\left(\frac{t}{k}\right)$ still belongs to E and $J_{\lambda/k}(\tilde{v}) = J_\lambda(v_\lambda)$. Furthermore, $J_{\lambda/k}(s\tilde{v}) = J_\lambda(sv_\lambda) \leqslant -s^2\frac{\lambda}{2}\int \langle v_\lambda, Lv_\lambda\rangle + s^\alpha \int G(v_\lambda)$ for $s \geqslant 1$, and since $\lambda \int \langle v_\lambda, Lv_\lambda\rangle = \int \langle v_\lambda, \nabla G(v_\lambda)\rangle \geqslant \int G(v_\lambda)$, we have $J_{\lambda/k}(s\tilde{v}) \leqslant \left(s^\alpha - \frac{s^2}{2}\right)\int G(v_\lambda) \leqslant 0$ if $s \geqslant 2^{1/2-\alpha}$. Thus $v_0 := 2^{1/2-\alpha}\tilde{v}$ can be connected to the endpoint e, defining Σ, with a path along which $J_{\lambda/k}$ is nonpositive, and we see that

$\text{minimax } J_{\lambda/k} \leqslant \max_{0 \leqslant s \leqslant 2^{1/2-\alpha}} J_{\lambda/k}(sv_0) = \max_{0 \leqslant s \leqslant 2^{1/2-\alpha}} J_\lambda(sv_\lambda)$.

Thus it is enough to show that :

$$\max_{0 \leqslant s \leqslant 2^{1/2-\alpha}} J_\lambda(sv_\lambda) \leqslant \alpha^{\alpha/(2-\alpha)} J_\lambda(v_\lambda)$$

To do this, first note that, with $\Lambda_\lambda(v) = -\lambda\langle v,Lv\rangle + \int \langle v,\nabla G(v)\rangle$, there results $s \frac{d}{ds} J_\lambda(sv) = \Lambda_\lambda(sv)$, and thus $J_\lambda(sv)$ is decreasing where $\Lambda_\lambda(sv)$ is negative. But, if $\Lambda_\lambda(v) = 0$ and $s \geqslant 1$, we have $\Lambda_\lambda(s,v) \leqslant -s^2 \int \langle v,\nabla G(v)\rangle + \alpha s^\alpha \int G(v) \leqslant (\alpha s^\alpha - s^2) \int G(v) \leqslant 0$ if $s \geqslant \alpha^{1/(2-\alpha)}$, and hence $\max_{0 \leqslant s \leqslant 2^{1/2-\alpha}} J_\lambda(sv_\lambda) \leqslant$ $\max_{0 \leqslant s \leqslant \alpha^{1/2-\alpha}} J_\lambda(sv_\lambda)$, because $\Lambda_\lambda(v_\lambda) = 0$. Now, if $s \geqslant 1$, we have, as above, $J_\lambda(sv_\lambda) \leqslant \left(s^\alpha - \frac{s^2}{2}\right) \int G(v_\lambda) \leqslant \alpha^{\alpha/2-\alpha} J_\lambda(v_\lambda)$ for $s \leqslant \alpha^{1/2-\alpha}$. On the other hand, if $s \leqslant 1$, then $\int G(sv_\lambda) \leqslant s \int G(v_\lambda)$ and so $J_\lambda(sv_\lambda) \leqslant s^2 J_\lambda(v_\lambda) + (s-s^2) \int G(v_\lambda) \leqslant [s^2 + \frac{2}{2-\alpha}(s-s^2)]\, J_\lambda(v_\lambda) \leqslant$ $\frac{1}{\alpha(2-\alpha)} J_\lambda(v_\lambda)$. Since $\frac{1}{\alpha(2-\alpha)} \leqslant \alpha^{\alpha/2-\alpha}$, (1.12) is proved.

But then, using the left inequality in (1.10) for minimax $J_{\lambda/k}$ and the right inequality in (1.10) for $J_\lambda(v_\lambda)$, we get

$$K^{\beta/\beta-2} \leqslant \alpha^{\beta/\beta-2} \left(\frac{M}{m}\right)^{\alpha/\beta-2} \quad \text{and so} \quad K \leqslant \alpha\left(\frac{M}{m}\right)^{1/\beta-1} .$$

Before describing the variational method which allows us to prove Theorem 1.6, we show how to derive Theorem 1.1, under an additional convexity hypothesis, from Theorem 1.4. We will use the same approximation argument as in [17].

Proof of Theorem 1.1 under a convexity hypothesis : Here we prove Theorem 1.1 under the additional hypothesis

(1.13) $\quad \exists h_o : \{(\xi,h) \in \text{epi } H : h > h_o\}$ is strictly convex

We can suppose $h_o = 1$, $\langle\nabla H(\xi),\xi\rangle \geqslant \beta H(\xi)$ if $H(\xi) \geqslant 1$, and $C := \{\xi \in \mathbb{R}^{2N} : H(\xi) \leqslant 1\}$ strictly convex. Let us define $\tilde{H}(\xi) = t^\gamma$ if $H(\xi/t) = 1$, with $2 < \gamma < \beta$, so that $\tilde{H}$ is a C^1 γ-homogeneous strictly convex function, and $\hat{\mathcal{H}}^{-1}(1) = \partial C$. Notice that $(H-\hat{\mathcal{H}})(t\xi) \geq t^{\beta}-t^{\gamma}$ for $t \geq 1$ and $H(\xi) = 1$. Denoting $T = \{\xi : 1 \leq H(\xi) \leq \hat{\mathcal{H}}(\xi)+1\}$, we define a new hamiltonian by setting $\hat{H}(\xi) = \hat{\mathcal{H}}(\xi)$ in C, $\hat{H}(\xi) = \hat{\mathcal{H}}(\xi) + \frac{1}{2}(H-\hat{\mathcal{H}})^2(\xi)$ in T, and $\hat{H}(\xi) = H(\xi) - \frac{1}{2}$ if $\xi \notin C \cup T$. It is easy to see that $\hat{H}$ is a C^1-strictly convex function, which coincides with $H - 1/2$ on the set $\{\hat{H} \geq h_1\}$, where $h_1 := \max\left\{\hat{\mathcal{H}}(\xi) + \frac{1}{2} : 1 \leq H(\xi) = \hat{\mathcal{H}}(\xi) + 1\right\}$. Thus it is enough to show that, for some $T_0 > 0$ and any $0 < T \leq T_0$, the system $z = \nabla\hat{H}(z)$ has a T-periodic solution Z_T satisfying $\hat{H}(z_T(t)) \equiv h_T > h_1$. To this extent, we first apply Theorem 1.4 (actually a slight improved version of it) to the Hamiltonian $\hat{H}_n(\xi) := \inf_{\eta} \{\hat{H}(\xi-\eta) + (\eta/\gamma)|\eta|^{\gamma}\}$ to get, for any T, a 2π-periodic solution $z_{n,T}$ of $\dot{z} = \frac{T}{2\pi} I\, \nabla\hat{H}_n(z)$. Since, if we denote by $\hat{G}_n$ the Legendre transform of $\hat{H}_n$, there results $\hat{G}_n \leq \hat{G}_1$, we infer, as in the proof of Theorem 1.4, a uniform bound for the energies $h_n \equiv \hat{H}_n(z_n,T)$, and hence for $\|\dot{z}_{n,T}\|_{L^\infty}$. So we can assume $z_{n,T}$ converges uniformly to some z_T. Since $\hat{H}_n$ converges uniformly, with its gradient, to $\hat{H}$ on compact sets, we conclude that z_T is a 2π-periodic solution of $\dot{z} = \frac{T}{2\pi}\, I\, \nabla\hat{H}(z)$. To see that $h_T \equiv \hat{H}(z_T) > h_1$ for T small, we argue by contradiction. Let $h_n \equiv \hat{H}(z_{T_n}) \leq h$ for a sequence $T_n \to 0$. Since $\nabla\hat{H}(\xi) \neq 0$ for $\xi \neq 0$, we see that $z_{T_n} \to 0$ uniformly. So z_{T_n} are 2π-periodic solutions for $\dot{z} = T_n/2\pi\, I\, \nabla\hat{\mathcal{H}}(z)$, with bounded energy. But this is clearly impossible, as one can see, for example, observing that the left inequality in (1.11) (and hence in (1.7) too) holds for the least critical value of J_λ, provided the Hamiltonian function is homogeneous.

We now go back to sketch the proof of Theorem 1.6. As already mentioned, it is a consequence of the following global result which will be proved in the next section.

Theorem 1.7 ([12]). Let $H \in C^2(\mathbb{R}^{2N},\mathbb{R})$ satisfy (1.5), (1.6). Assume furthermore

(1.14) $$\langle H''(\eta)\xi,\xi\rangle \geqslant c_1|\eta|^{\beta-2} \quad \forall\, \eta \in \mathbb{R}^{2N} \text{ and } |\xi| = 1 ,\ \xi \in \mathbb{R}^{2N}$$

(1.15) $$\langle G''(\eta)\eta,\eta\rangle < \langle \nabla G(\eta),\eta\rangle \qquad \forall\, \eta \in \mathbb{R}^{2N} ,\ \eta \neq 0 .$$

Then, for every $T > 0$, (H) has a periodic solution having T as minimal period.

In fact, if H is a Hamiltonian satisfying the assumptions of Theorem 1.6, it is easy to modify it, in such a way that the new hamiltonian coincides with the given one in a small neighborhood of zero and satisfies all the hypothesis of Theorem 1.7. Thus, in order to derive Theorem 1.6 from Theorem 1.7, it is enough to prove that, for T large, the T-periodic solutions, given by Theorem 1.7 when applied to the modified hamiltonian, have small energy. But this follows easily from energy estimates of the type established in Theorem 1.4 (see [12] for the details).

Remark 1.8. (i) A simple example of hamiltonian function which satisfies the assumptions of Theorem 1.7 is given by a C^2 β-homogeneous H, with $\beta > 2$. Nevertheless, a more direct constrained minimization argument on J_λ can be used in such a case.

(ii) Let us regard H as corresponding to a concave-convex Lagrangian $L(q,\dot{q})$, describing the motion of a mechanical system with N degrees of freedom ; in other words H is the Legendre transform of L with respect to the generalized velocities $\dot{q}$, while the dual function G is the Legendre transform of $-L$ with respect to the generalized coordinates q. Let L be given by some kinetic minus a potential energy, say $L(q,\dot{q}) = T(\dot{q}) - U(q)$; the assumptions of Theorem 1.7 hold if T is "subquadratic" while U is assumed β-homogeneous, with $\beta > 2$, since then $H = T^\star + U$ and $H^\star = T + U^\star$ ("$\star$" denoting the Legendre transform).

2. An abstract variational principle : minimal critical points.

In this section we describe an abstract variational principle which allows us to find a free critical point of a given (unbounded from below) functional as a minimum of a suitable constrained problem (see Hempel [21] in another context).

Let E be a real reflexive Banach space with norm $\|\cdot\|$, and dual space E'. The duality between E and E' will be denoted $<\cdot,\cdot>$. We consider a functional of the form :

$$f(u) = -\frac{1}{2} a(u,u) + b(u) \quad , \quad u \in E$$

where $a : E \times E \to \mathbb{R}$ and $b \in C^1(\mathbb{R},\mathbb{R})$ satisfy

(a_1) a is a continuous, symmetric bilinear form, positive somewhere.

(b_1) $\exists\, \alpha \in (1,2)$ and $m,M > 0 : m\|u\|^{\alpha} \leqslant <b'(u),u> \leqslant M\|u\|^{\alpha}$ $\forall\, u \in E$

(b_2) $\exists\, \theta < 2 : <b'(u),u> \leqslant \theta\, b(u) \qquad \forall\, u \in E$

(b_3) set $d(u) = <b'(u),u>$, then $d \in C^1(E,\mathbb{R})$ and $<d'(u,u)> < 2\, d(u)$

We remark that (a) to (b_3) are surely satisfied by the functional J_λ given by (1.9). And in fact, just as for J_λ , these properties, jointly with a P-S condition, are sufficient for the existence of a nontrivial critical point of saddle type of f . But here we are interested in additional properties of our critical point. More precisely we require that our critical point v satisfies the following "minimum-like" property :

$$(\star) \qquad -a(v,v) = \min \{-a(u,u) : b(su) = b(sv)\ ,\ \forall\, s \geqslant 0\ ,\ \langle b'(u),u\rangle = \langle b'(v),v\rangle\}\ .$$

Obviously, a minimum satisfies property ($\star$). To find such a stationary point for our (unbounded from below) functional, we introduce the constraint $M = \{u \in E \setminus \{0\} : h(u) = 0\}$, where $h(u) = \langle f'(u),u\rangle$. The idea is to prove that f achieves its minimum on M on a free critical point $\bar{v}$ of f , satisfying ($\star$). In particular $\bar{v}$ will minimize f on the set of its critical points. Henceforth we will be able to prove

Theorem 2.1. Let f be a weakly lower semicontinuous functional satisfying (a_1) to (b_3). Moreover, assume

(a_2) $\qquad A(u) = a(u,u)$ is (sequentially) weakly continuous.

Then f has a critical point satisfying ($\star$).

We begin with some lemmas. First note that every critical point of f belongs to M . Conversely, M is a "natural constraint" (see Berger [22]) for f , i.e. every critical point of $f|_M$ is a free critical point of f . We state this important fact as a lemma.

Lemma 2.2. Suppose $f,h \in C^1(E,\mathbb{R})$ is such that

$$(2.1) \qquad \langle h'(u),u\rangle \neq 0 \qquad \forall\, u \in M$$

Then $f'(u) = 0$, $u \neq 0$ iff u is a critical point of $f|_M$.

Notice that (2.1) implies M is a C^1 manifold. Again (2.1), and the Lagrange multiplier theorem, yields the result.

In view of (b_3), it is easily seen that Lemma 2.2 applies to our functional. We now list some properties of $f|_M$.

Lemma 2.3. Assume (a_1)-(b_1)-(b_3) . Then

(i) $\exists \rho > 0 : 0 < u < \rho$ implies $h(u) > 0$

(ii) $\exists c > 0 : f(u) \geqslant c\, u^2 \quad \forall\, u \in M$

(iii) $\frac{d}{ds}\left(h(su)/s^2\right) < 0 \quad \forall\, u \in E\, ,\ s > 0$

(iv) $\frac{d}{ds}\, f(su) = \frac{1}{s}\, h(su)$

In fact, (i) is a consequence of the subquadratic bahavior of b at zero, while (ii) is derived, using (b_1)-(b_3) and (i), from the form f takes on M : $f(u) = b'u) - \frac{1}{2} \langle b'(u),u \rangle \quad \forall\, u \in M$. Finally, a direct calculation and (b_3) imply (iii) and (iv).

Remark 2.4. Notice the geometrical meaning of Lemma 2.3. (i)-(ii) state that $tu \in M$, $\|u\| = 1$, for exactly one $t > 0$ and iff $a(u,u) > 0$. From (iv) we see that $f(tu)$, for a given u , is a strictly increasing function of t in (0,1) and strictly decreasing for $t \geqslant 1$.

Proof of Theorem 2.1 :

Step 1 : $f|_M$ achieves its minimum. By Lemma 2.3 (i)-(ii) we get $\inf_M f > 0$. Let $v_n \in M$ be a minimizing sequence, which, again by (ii), can be assumed weakly convergent to some v . Moreover, $v \neq 0$, for, otherwise, $a(v_n,v_n) = \langle b'(v_n),v_n \rangle \geqslant c_1 \|v_n\|^{\alpha}$ implies $\|v_n\| \to 0$, contradicting part (i) of Lemma 2.3. Now, observe that $a(v,v) > 0$ and so $\bar{v} = tv$ belongs to M for exactly one $t > 0$. Since by Lemma 2.3 (iv) $f(tv_n) \leqslant f(v_n)$, we get $f'tv) \leqslant \liminf f(tv_u) \leqslant \inf_M f$.

Step 2 : $\bar{v}$ satisfies (★) . In fact, if not, $-\frac{1}{2} a(\bar{v},v) > -\frac{1}{2} a(u,u)$ for some u for which $b(su) = b(s\bar{v}) \quad \forall\, s \geqslant 0$ and $\langle b'(u),u \rangle = \langle b'(\bar{v}),\bar{v} \rangle$. For such u , we have $h(u) < h(\bar{v}) = 0$, and so there is $s < 1$ such that $h(su) = 0$, by part (iii) of Lemma 2.3. But $f(su) = -\frac{1}{2} s^2 a(u,u) + b(su) < -\frac{1}{2} s^2 a(\bar{v},\bar{v}) + b(s\bar{v}) = f(s\,\bar{v}) < f(\bar{v})$, a contraidction.

We finish this section deriving Theorem 1.7 from Theorem 2.1. It is easy to see the J_λ given by (1.9) satisfies all the assumptions of

Theorem 2.1. It remains to prove that a point v satisfying (★) has minimal period 2π . If not, $u = v\left(\frac{t}{k}\right)$ is again in E for some $k > 1$. But $a(u,u) = \int_0^{2\pi} \langle u,Lu\rangle = k \int_0^{2\pi} \langle v,Lv\rangle > a(v,v)$, while $\int G(su) = \int G(sv)$, $\forall\, s \geqslant 0$ and $\int \langle \nabla G(u),u\rangle = \int \langle \nabla G(v),v\rangle$, contradicting (★).

III. A continuation method for second order hamiltonian systems.

A very powerful tool in nonlinear analysis is to get continuous branches of solutions for some "paramater dependent problem", in order to give existence results for prescribed values of the parameter. In our problem the parameter has a specific meaning : it is the period of the solution we are looking for, or its energy As already mentioned in the introduction, there are several striking results concerning continuous branches of periodic solutions emanating from an equilibrium. Roughly speaking, they state that three possible behaviors can occur along the branch : (i) the period goes to infinity ; (ii) the amplitude goes to infinity ; (iii) the orbit tends to some equilibrium and the corresponding period tends to the period of some periodic solution of the system linearized about such equilibria. Nevertheless, due to the difficulty of getting a priori estimates on the parameters involved, it seems hard to use such results to find solutions of prescribed period (or prescribed energy). To this extent, some more information is needed about the solution branch.

In a joint paper with Ambrosetti [12] we have considered the second order hamiltonian system

(3.1) $$\ddot{x} + \nabla U(x) = 0$$

We assumed on U

(U) $U \in C^2(\mathbb{R}^N, \mathbb{R})$ is strictly convex and $U(0) = 0$, $\nabla U(0) = 0$,

$U''(0)$ is a positive definite, symmetric matrix with

eigenvalues ω_j^2 , $0 < \omega_1^2 \leqslant \dots \leqslant \omega_N^2$.

Setting $V(x) = U(x) - \frac{1}{2}\langle U''(0)x,x\rangle$, and assuming it to be convex also, denote the Legendre transform by $\mathcal{V}$; we also require

(V_1) $\quad \exists\, \beta > 2$ and positive constants $c_1, c_2, \ldots$ such that :

$$c_1|\xi|^\beta \leqslant V(\xi) \leqslant c_2|\xi|^\beta \qquad \forall\, \xi \in \mathbb{R}^N$$

$$c_3|\eta|^{\beta-2} \leqslant \langle V''(\eta)\xi,\xi\rangle \qquad \forall\, \eta,\xi \in \mathbb{R}^N \text{ with } |\xi| = 1$$

$$0 < b\,V(\xi) \leqslant \langle V'(\xi),\xi\rangle \qquad \forall\, \xi \in \mathbb{R}^N$$

(V_2) $$\langle \mathcal{V}''(\xi)\xi,\xi\rangle < \langle \nabla \mathcal{V}(\xi),\xi\rangle \qquad \forall\, \xi \in \mathbb{R}^N \quad , \quad \xi \neq 0 .$$

In [12] we proved (see also Berger [21] for a local result)

Theorem 3.1. Assume $(U)-(V_1)-(V_2)$ and let $\omega = \frac{2\mu}{T} > \omega_N$. Then (3.1) has a solution v_ω having minimal period $T = \frac{2\pi}{T}$. Moreover, $\|v_\omega\|_{L^\infty} \to 0$ as $\omega \to \omega_N$ while $\|v_\omega\|_{L^\infty} \to \infty$ as $\omega \to \infty$.

The proof of Theorem 3.1 is based on the variational principle described in the preceding section. As well as for Theorem 1.7, one should like to drop the unpleasant assumption on the Legendre transform of the energy. We give here a result in this direction, but using continuation arguments.

First, we rewrite (3.1) as a nonlinear eigenvalue problem, looking for 2π-periodic solutions of

$$\omega^2\ddot{x} + \nabla U(x) = 0 \quad , \quad x \in C^2(\mathbb{R},\mathbb{R}^N) \tag{3.2}$$

Since double derivation is invariant under reflection in time, we are led to consider the two point boundary value problem

$$\begin{cases} \omega^2\ddot{x} + \nabla U(x) = 0 \\ \dot{x}(0) = \dot{x}(\pi) = 0 \end{cases} \tag{3.3}$$

If x is a solution for (3.3), we can extend it to $[-\pi,0]$ by evenness and then to all of $\mathbb{R}$ by periodicity, to get 2π-periodic solution of (3.2). We will look for increasing solutions of (3.3) ; the corresponding extensions will provide trajectories of (3.2) which are "increasing in a half period" and thus, surely, of minimal period 2π. Under rather strong positivity assumptions on U", we first show that there exists an unbounded connected set of increasing solutions (ω^2,x) (i.e. $\dot{x} \geqslant 0$) of (3.3) emanating from $(\omega_N^2,0)$; moreover the total energy is unbounded along the branch. Finally, we will see that, along the branch, $\omega^2 \to \infty$ as the energy tends to infinity.

Remark 3.2. Under much less restrictive assumptions on U, standard degree arguments yield the existence of a global branch of solutions of (3.3) emanating from $(\omega_N^2,0)$, and one could try to prove the energy goes to infinity along the branch and that the period is small at high energy. Actually, it is impossible, even in the framework of Theorem 3.1, to get a priori bounds on the minimal period. In fact, consider the system of uncoupled oscillators

$$\ddot{x}_i + f_i(x) = 0 \quad , \quad i = 1,2 \ , \quad \text{with} \quad f_i(0) = 0$$

where f_i are real convex functions growing more than linearly at infinity. It is easy to see that the orbit at the energy $\frac{1}{2}|\dot{x}|^2 + F_i(x_i) = c$ $(F_i' = f_i)$ of each uncoupled oscillator has (minimal) period tending to zero as c tends to infinity. Nevertheless, if x_i is a T_i-periodic solution (at high energy) of the i-th equation, and $T_1 T_2^{-1} = m\, n^{-1}$, then (x_1,x_2) has $n\,T_1 = m\,T_2$ minimal period, which can be arbitrarily large.

Theorem 3.3. Let $U \in C^2(\mathbb{R}^2,\mathbb{R})$ satisfy (U) and

(3.4) $$\frac{\partial^2 U}{\partial \xi_i\, \partial \xi_j}(\xi) \geqslant 0 \qquad \forall\, \xi \in \mathbb{R}^N \text{ and } i,j = 1,\dots,N \ ;$$

(3.5) $$\frac{\partial^2 U}{\partial \xi_j^2}(\xi) \geqslant \sigma(|\xi|)$$ where σ is such that $\sigma(\rho) \to +\infty$ as $\rho \to \infty$, and there exists $\delta > 0$: $\sigma(\rho) \geqslant \delta^2$, $\forall\, \rho \geqslant 0$.

Then, for every $T < \frac{2\pi}{\omega_N}$, (3.1) possesses a solution x_T having T as minimal period. Moreover $\|x_T\|_{L^\infty} \underset{T \to 0}{\to} \infty$.

Let us introduce some notation. Let $E = \{x \in C^1(0,\pi;\mathbb{R}^N) : \dot{x}(0) = \dot{x}(\pi) = 0\}$ with the usual norm $\|x\| = \|x\|_{L^\infty} + \|\dot{x}\|_{L^\infty}$, $K = \{x \in E : \dot{x} \geqslant 0\}$. K is a closed convex proper cone in E, with $E = K - K$. Let us denote by $\mathcal{S} \subset \mathbb{R}^+ \times (E \cap C^2)$ the set of nontrivial solutions of (3.2).

The first step in the proof of Theorem 3.3 is the following proposition, which actually does not involve the variational character of (3.3).

Proposition 3.4. Under the hypothesis of theorem 3.3, there exists a connected subset $\mathcal{S}^+ \subset \mathcal{S} \cap K$ such that $\sup\{\|x\|_{L^\infty} : (\omega^2, x) \in \mathcal{S}^+\} = \infty$ and $(\omega_N^2, 0) \in \overline{\mathcal{S}^+}$.

Proof of the Proposition : Due to the presence of a nontrivial kernel in the linear part of (3.3) we need first a Lyapunov-Schmidt reduction. Denote $E_o = \left\{u \in E : \int_o^\pi u = 0\right\}$. Every $x \in E$ can be written in a unique way as $x = u + \xi$, $u \in E_o$, $\xi \in \mathbb{R}^N$. Clearly, $x = u + \xi$ is a solution of (3.3) iff :

$$(3.6)_a \qquad \int_o^\pi \nabla U(u+\xi) = 0$$

$$(3.6)_b \qquad \omega^2 \ddot{u} + \nabla U(u+\xi) = 0$$

Because of the strict convexity of U, $(3.6)_a$ has a unique solution $\xi(u)$ foe every $u \in E_o$; moreover ξ depends in a C^1 fashion on u by the implicit function theorem.

We now study $(3.6)_b$, setting $\xi = \xi(u)$. Denote by Lh, $h \in E_o$, the unique solution in E_o of $-\ddot{u} = h$. Thus $(3.6)_b$ is equivalent to

(3.7) $\omega^2 u = L\ \nabla U(U+\xi(u))$

Set $N(u) = L\ \nabla U(u+\xi(u))$; notice that $N(K_o) \subset K_o$, where $K_o = K \cap E_o$. In fact $v = N(u)$ means $-\ddot{u} = \nabla U(u+\xi(u))$ and so $y = \dot{v}$ satisfies

$$\begin{cases} -\ddot{y} = U''(u+\xi(u))\dot{u} \\ y(0) = y(\pi) = 0 \end{cases}$$

Since $u \geqslant 0$, we infer by (3.4) that $y \geqslant$ .e. $v \in K_o$. In order to apply the Krein-Rutman theory to the bifurcation analysis of $(3.6)_b$ (see [23], [24]), we need to calculate the spectral radius of $N'(0)$. Since the largest μ for which $-\mu\ddot{u} = U''(0)u$ has a nontrivial solution in E_o is ω_N^2 , the usual bifurcation theory for compact cone preserving maps yields the existence of an unbounded connected set of solutions $S_o^+ \subset \mathbb{R}^+ \times K_o$ of (3.7), emaning from $(\omega_N^2, 0)$. Thus $S^+ = \{(\omega^2, u + \xi(u)) : (\omega^2, u) \in S_o^+\}$ is connected and unbounded in $\mathbb{R} \times K$.

To complete the proof, we first show that $\inf\{\omega^2 : (\omega^2, u) \in S_o^+\} \geqslant \delta^2$. In fact, if $(\omega^2, u) \in S_o^+$, $y = \dot{u}$ satisfies

$$\begin{cases} \omega^2\ddot{y} + U''(u+\xi(u))y = 0 \\ y(0) = y(\pi) = 0 \end{cases}$$

Thus, by (3.4) and (3.5), $0 \geqslant \omega^2 y_j + (\partial^2 U / \partial\xi_j^2) y_j \geqslant \omega^2 y_j + \delta^2 y_j$ for $j = 1,\ldots,N$. Since for some j it results $y_j > 0$, this readily implies $\omega^2 \geqslant \delta^2$.

Using the bound on ω^2 , we can now show that $\sup\left\{\|x\|_{L^\infty} : (\omega^2, x) \in S^+\right\} = +\infty$.

In fact, if not, from (3.3) and the lower bound on ω^2, we infer the boundedness of $\|x\|$ along the branch and thus the existence of a sequence $(\omega_n^2, x_n) \in S^+$ with $\omega_n^2 \to \infty$. But this is clearly impossible, since, setting $y_n = \dot{x}_n$, it should result $y_n + \left(U''(x_n)/\omega_n^2\right) y_n = 0$,

$y_n(0) = y_n(\pi) = 0$, while this linear system has no nontrivial solution for ω_n^2 large enough, since $U''(x_n)$ is uniformly bounded.

<u>Proof of Theorem 3.3 completed :</u> By Proposition 3.4 we can choose a sequence $(\omega_k^2, x_k) \in S^+$ with $\|x_k\| \to \infty$. Let us show that $\omega_k^2 \to \infty$. We first give some existence, and omit the subscript k here.

Let $(\omega^2,x) \in S^+$ and $z(t) = x(\omega t)$, $t \in [0,\frac{\pi}{\lambda}]$, be the corresponding "increasing" trajectory of (3.1) (the closed orbit arises describing again the same trajectory but in opposite direction). Along z the total energy remains constant : $\frac{1}{2}\omega^2 |\dot{x}(t)|^2 + U(x(t)) \equiv U(x(0)) = \mathcal{E}_x$; thus the trajectory lies in the region $\{U(\xi) \leqslant \mathcal{E}_x\}$. Remark that, since $\omega^2 \geqslant \delta^2$ along the branch , $\sup_{S+} \mathcal{E}_x = +\infty$; in particular $\sup_{S+} |x(0)| = +\infty$. The length of the trajectory $\ell_x = \int_0^\pi |\dot{x}(\tau)|d\tau$, is related to ℓ_x by the relation

$$(3.8) \qquad \frac{\ell_x}{\sqrt{\mathcal{E}_x}} \underset{\|x\| \to \infty}{\to} 0$$

In fact, since x is increasing, ℓ_x is less than N times the diameter of the region $\{U(\xi) \leqslant \mathcal{E}_x\}$. Thus (3.8) follows from (3.5), since this implies $U(\xi)|\xi|^{-2} \to 0$ as $|\xi| \to \infty$.

We now relate the time x spends at "low" energy, with the frequency ω . For this, let us denote $F_x = \{t \in [0,\pi] : U(x(t)) \leqslant \mathcal{E}_x/2\}$. From the energy conversation law we get $\frac{1}{2}\omega^2 |\dot{x}(t)|^2 \geqslant \mathcal{E}_x/2 \quad \forall\, t \in F_x$, and thus $\ell_x \geqslant \int_{F_x} |\dot{x}| \geqslant (\sqrt{\mathcal{E}_x}/\omega)\ |F_x|$, i.e.

$$(3.9) \qquad |F_x| \leqslant \frac{\ell_x}{\sqrt{\mathcal{E}_x}}\ \omega$$

Finally, since, as already noticed, we have $\omega^2 y_j + (\partial^2 U/\partial\xi_j^2) y_j \leqslant 0$ and $y_j(0) = y_j(\pi) = 0$ for $j = 1,\ldots,N$, where $y = \dot{x}$, we obtain $1 \leqslant \lambda_1(\phi_j)$, where $\lambda_1(\phi_j)$ denotes the first eigenvalue of the problem $\ddot{v} + \phi_j(t)v = 0$, $v(0) = v(\pi) = 0$ and $\phi_j(t) = \omega^{-2}(\partial^2 U/\partial\xi_j^2)\ (x(t))$.

But, setting $O_x = [0,\pi] \setminus F_x$, we have $\phi_j \geq \omega^{-2}\sigma(\rho_x)\chi_x$, if $\rho_x = \inf\{|\xi| : U(\xi) \geq \mathcal{E}_x/2\}$ and χ_x denotes the characteristic function of O_x. Thus $1 \leq \omega^2 \lambda_1(\chi_x)(\sigma(\rho_x))^{-1}$, that is

$$(3.10) \qquad \omega^2 \geq \sigma_1(\rho_x)\,\lambda_1^{-1}(\chi_x)$$

Now, the variational characterization of $\lambda_1(\chi_x)$ implies $\lambda_1^{-1}(\chi_x) \geq \int_{O_x} \psi^2 / \int_0^\pi \dot\psi^2 \quad \forall\, \psi \in H_o^1$, and so $\lambda_1^{-1}(\chi_x) \geq \frac{2}{\pi} \int_{O_x} \sin^2 t = 1 - \frac{2}{\pi} \int_{F_x} \sin^2 t \geq 1 - \frac{2}{\pi}|F_x|$. Henceforth, substituting in (3.10) and using (3.9) we get

$$(3.11) \qquad \omega^2 \geq \sigma(\rho_x)\left(1 - \frac{2}{\pi}\,\frac{\ell_x}{\sqrt{\mathcal{E}_x}}\,\omega\right)$$

which implies $\omega_n \to \infty$ for $\|x_n\| \to \infty$ in view of (3.8) and the growth assumption (3.5).

Remark 3.5. From Theorem 3.3 it is easy to obtain the existence of solutions of arbitrarily long (minimal) period providing U is superquadratic in zero : $U''(0) = 0$. In fact, it is enough to consider an approximating system

$$(3.12) \qquad \lambda^2\ddot{x} + \varepsilon_n^2 x + U'(x) = 0 \quad , \quad \varepsilon_n \to 0$$

for which a solution x_n having 2π as minimal period exists, if n is sufficiently large, in view of Theorem 3.3. Since the estimates (3.8) and (3.11) still obviously hold, $\mathcal{E}_n \equiv \frac{1}{2}|\dot{x}_n|^2 + \frac{\varepsilon_n^2}{2}|x_n|^2 + U(x_n)$ is a bounded sequence, and hence $\sup_n \|x_n\|_{L^\infty} < +\infty$. Thus x_n has a subsequence which converges to a solution x of (3.12) with $\varepsilon = 0$. Since such a solution still belongs to K, $x(\omega t)$ is a solution of (3.1) of minimal period $2\pi/\omega$.

REFERENCES

[1] Weinstein, A., Normal modes for nonlinear Hamiltonian systems, Inv. Math. 20 (1973), 47-57.

[2] Moser, J., Periodic orbits near an equilibrium and a theorem by Alan Weinstein, Comm. Pure Appl. Mth. 29 (1976) 727-747.

[3] Lyapunov, A., Problème général de la stabilité du mouvement, Ann. Fac. Sci. Toulouse (2) (1907) 203-474.

[4] Fadell, E.R., and P.H. Rabinowitz : Generalized cohomological index theories for Lie group actions with an application to bifurcation questions for Hamiltonian systems, Inv. Math. 45 (1978), 139-174.

[5] Ekeland, I., and J.M. Lasry : On the number of periodic trajectories for a Hamiltonian flow on a convex energy surface, Ann. of Math. 112 (1980), 283-319.

[6] Ambrosetti, A. and G. Mancini : On a theorem by Ekeland and Lasry concerning the number of periodic hamiltonian trajectories, J. Diff. Equat. (in press).

[7] Amann, H. and E. Zehnder : Multiple periodic solutions of asymptotically linear Hamiltonian equations,

[8] Rabinowitz, P.H. : Periodic solutions of Hamiltonian systems, Comm. Pure Appl. Math. 31 (1978) 157-184.

[9] Rabinowitz, P.H. : A variational method for finding periodic solutions of differential equations, Nonlinear evolution equations (M.G. Crandall, ed.) Academic Press (1928) 225-251.

[10] Benci, V. and P.H. Rabinowitz : Critical point theorem for indefinite functionals, Inv. Math. 52 (1979) 336-352.

[11] Clarke, F. and I. Ekeland : Hamiltonian trajectories having prescribed minimal period, Comm. Pure Appl. Math. 33 (1980) 103-116.

[12] Ambrosetti, A. and G. Mancini : Solutions of minimal period for a class of convex Hamiltonian systems, Math. Ann. 255 (1981) 405-421.

[13] Alexander, J.C. and J.A. Yorke : Global bifurcation of periodic orbits, Amer. J. Math. 100 (1978) 263-292.

[14] Yorke, J.A. and J. Mollet-Peret, preprint.

[15] Berger, M.S., Nonlinearity and functional analysis, Academic Press, New-York, 1978.

[16] Rabinowitz, P.H. : Periodic solutions of Hamiltonian systems : a survey, preprint.

[17] Ekeland, I., Periodic solutions of Hamiltonian equations and a theorem of P. Rabinowitz, J. Diff. Equat. 34 (1979) 523-534.

[18] Ambrosetti, A. and P.H. Rabinowitz : Dual variational methods in critical point theory and applications, J. Funct. Anal. 14 (1973) 349-381.

[19] Clarke, F., Periodic solutions to Hamiltonian inclusions, J. Diff. Equat. 40 (1981), 1-7.

[20] Brezis, H., J.M. Coron and L. Nirenberg : Free vibrations for a nonlinear wave equation and a theorem of P. Rabinowitz, Comm. Pure Appl. Math. 33 (1980) 667-684.

[21] Hempel, J.A., Multiple solutions for a class of nonlinear boundary value problems, Ind. Univ. Math. J. 20 (1971), 983-996.

[22] Berger, M.S., Critical point theory for nonlinear eigenvalue problems with indefinite principal part, Trans. Amer. Math. Soc. 186 (1973) 151-169.

[23] Amann, H., Fixed point equation and nonlinear eigenvalue problems in ordered Banach spaces, SIAM Review 18 (1976) 620-709.

[24] Dancer, C.N., Solution branches for mappings in cones, and applications, Bull. Austral. Math. Soc. 11 (1976) 131-143.

DUALITY IN NON CONVEX VARIATIONAL PROBLEMS

I. EKELAND - J.M. LASRY

CEREMADE.

DUALITY IN NON CONVEX VARIATIONAL PROBLEMS

I. EKELAND - J.M. LASRY

CEREMADE

RESUME : On donne une formulation abstraite d'une méthode de dualité permettant de trouver les points critiques de certaines fonctionnelles comprenant un terme quadratique et un terme convexe. On donne deux applications l'une aux systèmes différentiels hamiltoniens, l'autre à une variante de l'équation des ondes.

ABSTRACT: We give an abstract formulation of a duality method which enables us to find critical points of certain functionals which split into a quadratic term and a convex term. Two applications are given, to hamiltonian differential systems and to a variant of the wave equation.

Key words : Duality, critical point, non convex optimisation wave equation, hamiltonian problems.

Code AMS : 47 H 15 49 H 10 .

0. INTRODUCTION

Duality theory now is a standard tool for dealing with convex optimization problems (see [15], [16]). Basically, one uses the Legendre transform to associate with any given problem (the primal) another one (the dual), in such a way that the solution of the second problem trivializes the first.

The first attempt to extend this method beyond convexity is due to J. TOLAND ([17], [18]). His works deals with optimizing functions which can be written as F - G , with F and G both convex.

On another track, J.P. AUBIN and I. EKELAND ([4]) introduced duality theory in the study of Hamiltonian, or, more generally, hyperbolic problems. This kind of problem exhibits a particular kind of nonlinearity. The right formulation was found by F. CLARKE ([8]; see also CLARKE - EKELAND [9]) and has met with considerable success ([10]; [11]; [12], [13], [14]) in the study of Hamiltonian systems.

H. BREZIS, J.M. CORON, and L. NIRENBERG have taken up this method, and extended these results to a non-linear wave equation ([5], [6]). Others have also started working in this direction ([1], [19]).

It is the purpose of this paper to present an abstract formulation of the CLARKE - EKELAND duality theory which will cover all these results, and to illustrate it by new ones. The two first sections are devoted to the general (abstract) result, the two last ones to applications.

I. ABSTRACT DUALITY THEOREMS.

Let V be a reflexive Banach space over R, with dual $V^\star$. We are given a continuous quadratic form Q over V. This means that $Q : V \to R$ can be written as $Q(x) = B(x,x)$ with $B : V \times V \to R$ a continuous bilinear functional. It is well-known that Q can be written in a unique way as :

$$Q(x) = \frac{1}{2} \langle Ax,x \rangle$$

with $A : V \to V^\star$ a continuous linear map such that :

$$A = A^\star \quad .$$

We shall call A <u>self-adjoint</u>. Note that we do not require A to be injective nor subjective, and that Q can very well change signs, or even vanish on a subspace. It will not be convex, unless it stays non-negative.

We will also be given a convex lower semi-continuous function $F : V \to R \cup \{+\infty\}$, not identically $+ \infty$. Recall (from [25] for instance) that its subgradient $\partial F(\bar{u})$ at $\bar{u} \in V$, and its conjugate $F^\star : V^\star \to R \cup \{+\infty\}$ are defined by :

$$\partial F(\bar{u}) = \{u^\star \in V^\star \mid F(u) \geqslant F(\bar{}) + \langle u^\star, u-\bar{u} \rangle \ , \ \forall \ u \in V\}$$

$$F^\star(u^\star) = \sup \ \{\langle u^\star,u \rangle - F(u) \mid u \in V\}$$

and that we have Fenchel's reciprocity formula :

$$u^\star \in \partial F(u) \ \Leftrightarrow \ u \in \partial F^\star(u^\star) \ .$$

DEFINITION 1 : A critical point of the function

$$I(u) = \frac{1}{2} \langle Au,u\rangle + F(u)$$

is a point $\bar{u} \in V$ where :

$$- A\bar{u} \in \partial F(\bar{u}) \qquad ./$$

This reduces to the usual definition when F is differentiable. In the general, non-smooth, case we have the following.

PROPOSITION 2 : Assume that the restriction of I to any straight line running through $\bar{u}$ has a local minimum or a local maximum at $\bar{u}$. Then $\bar{u}$ is a critical point.

Proof : Assume u_o is not a critical point : that is, $-Au_o$ does not belong to $\partial F(u_o)$. This means that there exists some point $u_1 \in V$ and some $\varepsilon > 0$ such that :

$$F(u_1) \leqslant F(u_o) - \langle Au_o , u_1 - u_o\rangle - \varepsilon$$

Set $u_t = tu_1 + (1-t)u_o \in V$. The function $t \to F(u_t)$ defined on the real line is convex. It follows that the slope $t^{-1}[F(u_t) - F(u_o)]$ is a decreasing function of t, and hence, for all $t < 1$ and $\neq 0$:

$$t^{-1} [F(u_t)-F(u_o)] \leqslant F(u_1)-F(u_o) \leqslant - \langle Au_o,u_1-u_o\rangle - \varepsilon$$

Setting $\langle A(u_1-u_o) , u_1-u_o\rangle = a$, we have also :

$$\frac{1}{2} \langle Au_t,u_t\rangle = \frac{1}{2} \langle Au_o,u_o\rangle + t \langle Au_o,u_1-u_o\rangle + \frac{t^2}{2} a$$

Combining the last two relations, we get :

$$0 < t < 1 \Rightarrow F(u_t) + \frac{1}{2} \langle Au_t,u_t\rangle < F(u_o) + \frac{1}{2} \langle Au_o,u_o\rangle -\varepsilon t + \frac{t^2}{2} a$$

$$t < 0 \qquad \Rightarrow F(u_t) + \frac{1}{2} \langle Au_t,u_t\rangle > F(u_o) + \frac{1}{2} \langle Au_o,u_o\rangle -\varepsilon t + \frac{t^2}{2} a$$

If we choose $|t| < 2\varepsilon|a|^{-1}$, then $-\varepsilon t + at^2/2$ has a same sign as $-\varepsilon t$, negative in the first case and positive in the second. This proves that $I(u_t) = F(u_t) + \langle Au_t,u_t\rangle /2$ can have neither a local maximum nor a local minimum at $t = 0$ ./

Corollary 3 : If I has a local minimum or a local maximum at $\bar{u}$, then $\bar{u}$ is a critical point ./

We now give the main result of this paper. It is a duality result, in the spirit of convex analysis (see [15], [16]), although we are dealing with a non-convex problem. We denote by R(A) the range of A and by Dom $F^\star$ the domain of $F^\star$, i.e., the set of all points $v^\star \in V^\star$ where $F^\star(v^\star) < +\infty$.

THEOREM 4.

Define two functions I and J on V by

$$I(u) = \frac{1}{2} \langle Au,u\rangle + F(u)$$

$$J(v) = \frac{1}{2} \langle Av,v\rangle + F^\star(-Av) .$$

Any critical point of I is also a critical point of J. If

$$0 \in \text{Int } (\text{Dom}^\star + R(A))$$

then a converse holds : if $\bar{v}$ is a critical point of J, there is some $\bar{w} \in \text{Ker } A$ such that $\bar{v} + \bar{w}$ is a critical point of I ./

Proof : Before starting, let us write down the condition for $\bar{v}$ to be a critical point of J. From the definition we get :

$$- A\bar{v} \in \partial [F^\star \circ (-A)] \quad (\bar{v})$$

It is a result of Aubin (see [3]) that for any continuous linear operator $B : V \to V^\star$ we have :

$$\partial(F^\star \circ B)(\bar{v}) \supset B^\star \partial F^\star (B\bar{v})$$

and that equality holds provided $0 \in \text{Int}\,(\text{Dom}\,F^\star - R(B))$. We shall use this result, with $B = -A = B^\star$.

Let $\bar{u}$ be a critical point of I. By definition, we have $-A\bar{u} \in \partial F(\bar{u})$. By Fenchel's reciprocity formula, this can also be written $\bar{u} \in \partial F^\star(-A\bar{u})$. Applying $-A$ to both sides, we get :

$$- A\bar{u} \in - A\partial F^\star(- A\bar{u})$$

Since $- A\partial F^\star(- A\bar{u})$ is contained in $\partial\,[F^\star \circ (-A)]\,(\bar{u})$, this implies that $\bar{u}$ is a critical point of J.

Conversely, assume $0 \in \text{Int}\,(\text{Dom}^\star + R(A))$, and let $\bar{v}$ be a critical point of J. By Aubin's result just quoted, we have :

$$- A\bar{v} \in - A\partial F^\star(- A\bar{v})$$

This means that there is some $\bar{w} \in \text{Ker}\,A$ such that :

$$\bar{v} + \bar{w} \in \partial F^\star(- A\bar{v})$$

Set $\bar{u} = \bar{v} + \bar{w}$. We have $A\bar{v} = A\bar{u}$, since $A\bar{w} = 0$, and the equation can be written :

$$\bar{u} \in \partial F^\star(- A\bar{u})$$

We now use Fenchel's reciprocity formula once more, to get $- A\bar{u} \in \partial F(\bar{u})$, the desired result ./

In applications of this result to functional analysis, where V usually is some Sobolev space and $V^\star$ the corresponding space of distributions, the condition that the interior of $\text{Dom}\,F^\star + R(A)$ in $V^\star$ must contain the origin is very hard to satisfy. On the other hand, F usually factorizes through some appropriate L^p space. We therefore give a variant of theorem 4, more suitable for applications to functional analysis.

THEOREM 5.

We are given another reflexive Banach space X, a l.s.c. convex function $G : X \to \mathbb{R} \cup \{+\infty\}$, with $\text{Dom}\, G \neq \emptyset$, and a continuous linear map $K : V \to X$. Assume that the map $v \to (Av, Kv)$ has closed range in $V^\star \times X$.

Consider the linear subspace $M = (K^\star)^{-1} R(A)$ of $X^\star$, and assume that :

$$0 \in \text{Int}\,(\text{Dom}\, G^\star - M) \subset X^\star$$

Define two functions $I : V \to \mathbb{R} \cup \{+\infty\}$ and $J : V \times X^\star \to \mathbb{R} \cup \{+\infty\}$ by :

$$I(u) = \frac{1}{2} \langle Au, u\rangle + G(Ku)$$

$$J(v, x^\star) = \frac{1}{2} \langle Av, v\rangle + G^\star(x^\star)$$

Define in $V \times X^\star$ a closed linear subspace L by the equation $Av + K^\star x^\star = 0$, and denote by J_L the restriction of J to L.

If J_L has a critical point $(\bar{v}, \bar{x}^\star) \in L$, then there is some $\bar{w} \in \text{Ker}\, A$ such that $\bar{u} = \bar{v} + \bar{w}$ is a critical point of I.

Proof. Let us first clarify the condition that $(\bar{v}, \bar{x}^\star)$ be a critical point of J_L. Note first that $(\bar{v}, \bar{x}^\star)$ must belong to L, and hence :

$$A\bar{v} + K^\star \bar{x}^\star = 0$$

Define $\overrightarrow{G^\star} : V \times X^\star \to \mathbb{R} \cup \{+\infty\}$ by $\overrightarrow{G^\star}(v, x^\star) = G^\star(x^\star)$. Denote by $i : L \to V \times X^\star$ the standard injection. We have :

$$\partial(\overrightarrow{G^\star} \circ i)(v, x^\star) = i^\star \partial\overline{G} \circ i\ (v, x^\star)$$

provided Aubin's condition $0 \in \text{Int}\,(\text{Dom}\, \overrightarrow{G^\star} - R(i))$ in $V \times X^\star$ is satisfied. Since $\text{Dom}\, \overrightarrow{G^\star} = V \times \text{Dom}\, G^\star$, this boils down to $0 \in \text{Int}\,(\text{Dom}\, G^\star - M)$, which is satisfied. Relating $\overrightarrow{G^\star}$ to $G^\star$, we finally get :

$$\partial(\overrightarrow{G^\star} \circ i)(v, x^\star) = i^\star\, [0, \partial G^\star(x^\star)]$$

The condition that $(\bar{v},\bar{x}^\star)$ be a critical point of J_L over L can be written :

$$i^\star([-A\bar{v}, 0] - [0, \partial G^\star(\bar{x}^\star)]) = 0$$

We have to look at the kernel of $i^\star$. This is the set of all continuous linear functionals on $V \times X^\star$ which vanish on L. Since L is defined by the equation $Av + K^\star x^\star = 0$, we find $\operatorname{Ker} i^\star$ to be the closure of the set of all pairs (Au,Ku) in $V^\star \times X$ when u (we will take -u for the sake of convenience) runs over V. By assumption, this set is already closed, so that we can write :

$$\bar{u} \in V \ : \ [-A\bar{v},0] \in [0,\partial G^\star(\bar{x}^\star)] - [A\bar{u},K\bar{u}]$$

This splits into $A(\bar{u}-\bar{v}) = 0$ and $K\bar{u} \in \partial G^\star(\bar{x}^\star)$. The first equation can also be written $-A\bar{u} = K^\star\bar{x}^\star$.

Using Fenchel's reciprocity formula on $K\bar{u} \in \partial G^\star(\bar{x}^\star)$, we get $\bar{x}^\star \in \partial G(K\bar{u})$. Applying $K^\star$ to both sides, we get :

$$K^\star\bar{x}^\star = -A\bar{u} \in K^\star\partial G(K\bar{u})$$

But the right-hand side is known to be contained in $\partial(G \circ K)$, so that $-A\bar{u} \in \partial(G \circ K)(\bar{u})$. The result is proved ./

We still have the condition that the set $\{(Av,Kv) \mid v \in V\}$ be closed in $V^\star \times X$, which can be troublesome for some applications. We get around it by one more sophistication :

<u>THEOREM 6.</u>

Let V and X be reflexive Banach spaces, $G : X \to \mathbb{R} \cup \{+\infty\}$ a l.s.c. convex function with $\operatorname{Dom} G \neq \emptyset$, and $K : V \to X$ a continuous linear map. Define in $X^\star$ a linear subspace $M = \{x^\star \mid \ v \in V : Av = K^\star x^\star\}$, and assume that :

$$0 \in \operatorname{Int}(\operatorname{Dom} G^\star - M) \subset X^\star$$

Assume moreover that there is a map $S \in \mathcal{L}(M,V)$ such that

$$K^{\star}x^{\star} + ASx^{\star} = 0 \qquad \text{all } x^{\star} \in M$$

This will imply that M is closed. Now consider the two functions $I : V \to R \cup \{+\infty\}$ and $\overline{J} : M \to R \cup \{+\infty\}$ defined by :

$$I(u) = \frac{1}{2} \langle Au,u\rangle + G(Ku)$$

$$\overline{J}(x^{\star}) = -\frac{1}{2} \langle KSx^{\star},x^{\star}\rangle + G^{\star}(x^{\star})$$

If $\overline{J}$ has a critical point $\overline{x}^{\star}$ over M, then there is some $\overline{y} \in M^{\perp}$ such that, setting $S\overline{x}^{\star} = \overline{v}$,

$$- A\overline{v} \in K^{\star}\partial G(K\overline{v} + \overline{y}) \qquad /$$

Proof : Let us first check that M is closed. Since S is a bounded linear map, it can be extended to the closure $\overline{M}$ of M. We thus have a map $\overline{S} \in L(\overline{M},V)$, and by continuity $K^{\star}x^{\star} + ASx^{\star} = 0$ for all $x^{\star} \in \overline{M}$. But this implies that $x^{\star} \in M$, and hence $M = \overline{M}$.

Denote by $i = M \to X^{\star}$ the injection. The transpose of $K^{\star} \circ i$ is $i^{\star} \circ K$. From the equation $K^{\star} \circ i + AS = 0$ in $L(M,V)$ it follows easily that :

$$S^{\star}K^{\star} \circ i = - S^{\star}AS = (i^{\star} \circ K)\, S$$

Now let $\overline{x}^{\star}$ be a critical point of $\overline{J}$ over M. By definition we have

$$(i^{\star} \circ K)\, S\, \overline{x}^{\star} \in \partial(G^{\star} \circ i)(\overline{x}^{\star})$$

Since Aubin's condition $0 \in \text{Int}\,(\text{Dom}\, G^{\star} - M)$ is satisfied, we know that the subdifferential on the right is $i^{\star} \circ \partial G^{\star}(\overline{x}^{\star})$. The equation thus becomes :

$$i^{\star}\, [KS\overline{x}^{\star} - \partial G^{\star}(\overline{x}^{\star})] \supset 0$$

Hence, there exists some $\overline{y} \in \text{Ker}\, i^{\star} = M^{\perp}$ such that :

$$KS\overline{x}^{\star} + \overline{y} \in \partial G^{\star}(\overline{x}^{\star})$$

The result follows by using Fenchel's identity :

$$\bar{x}^{\star} \in \partial G(KS\bar{x}^{\star} + \bar{y})$$

and applying $K^{\star}$ to both sides ./

To understand the meaning of theorem 6, note that $M^{\perp}$ always contains $\overline{K(\text{Ker } A)}$, and M actually coincides with $\overline{K(\text{Ker } A)}$ whenever A has closed range in $V^{\star}$. Indeed, in this case $R(A) = (\text{Ker } A)^{\perp}$ (remember A is self-adjoint), so that M is characterized by the equation $\langle K^{\star}x^{\star}, v\rangle = 0$ for all $v \in \text{Ker } A$.

Going one step further, if $\bar{y}$ actually belongs to K(Ker A) itself, $\bar{y} = K\bar{w}$ with $A\bar{w} = 0$, then $\bar{u} = S\bar{x}^{\star} + \bar{w}$ is a solution of $\bar{u} \in K^{\star}\partial G(K\bar{u})$ and hence a critical point of I. This motivates the following :

<u>DEFINITION 7</u> : Any point $\bar{x} \in X$ which splits into $\bar{x} = K\bar{v} + \bar{y}$ with $A\bar{v} + K^{\star}\partial G(\bar{x}) \supset 0$ and $\bar{y} \in M^{\perp}$ is called a pseudo-critical point of I./

Note that if R(A) is closed in $V^{\star}$ and K(Ker A) is closed in X, then any pseudo-critical point is critical point.

II. ABSTRACT EXISTENCE THEOREMS

The reason the duality theorems of the preceding section are of interest is that it may be much easier to find critical points for J, or J_L, or $\bar{J}$, than for I itself. For instance, these dual functionals may well have a global minimum while I is unbounded above and below. We illustrate this with several existence results. The general setting is as described in the preceding section, particularly theorem 1-6. We classify our results according to the growth assumptions we make on $G : X \to R \cup \{+\infty\}$.

A - SUBQUADRATIC G

PROPOSITION 1 : Assumptions as in theorem I-6. Assume moreover K is a compact map, and there are constants $k > \|KS\|$ and $c \in R$ such that

$$\forall\, x \in X \quad , \quad G(x) \leqslant \frac{1}{2k} \|x\|^2 + c$$

Then I has a pseudo-critical point $\bar{x} \in X$.

Proof : It is sufficient to prove that $\bar{J}$ has a global minimum over M. Recall that M is closed. First we seek an a priori estimate. Using the definition of $G^\star$ and the condition on G, we get :

$$G^\star(x^\star) \geqslant \frac{k}{2} \|x^\star\|^2 - c$$

Writing this into $\bar{J}$, we get :

$$\bar{J}(x^\star) \geqslant \frac{1}{2} (k - \|KS\|) \|x^\star\|^2 - c$$

It follows that any minimizing sequence is bounded. Choose a minimizing sequence $x^\star_M$, weakly converging to $\bar{x}^\star$

$$\bar{J}(x^\star_M) \to \operatorname*{Inf}_M \bar{J}(x^\star)$$

Since $G^\star$ is convex and l.s.c., it is weakly l.s.c. :

$$\liminf_n G^\star(x^\star_n) \geqslant G^\star(\bar{x}^\star)$$

Since K is compact, so is KS. It follows that the first term in $\bar{J}$ is weakly continuous on bounded sets :

$$\lim_n \langle x^\star_n , KSx^\star_n\rangle = \langle \bar{x}^\star , KS\bar{x}^\star\rangle$$

We finally get $\bar{J}(\bar{x}^\star) \leqslant \lim_n \bar{J}(\bar{x}^\star_n)$, so that $\bar{x}^\star$ minimizes $\bar{J}$ on M./

It sometimes happens that one is interested in finding <u>non-trivial</u> critical points. The first requirement for this is that there already be a trivial solution. This occurs when G is minimum at the origin :

$$\forall\, x \in X \qquad G(x) \geqslant G(0)$$

so that zero is an obvious critical point of I. The value of the minimum G(0) is usually taken to be zero, since adding a constant to G will not change the critical points of I.

<u>PROPOSITION 2</u> : Assumptions as in theorem 1-6. Assume moreover that K is compact and that G satisfies the following, for some $k > \|KS\|$ and $c \in R$:

$$\forall\, x \in X \qquad 0 = G(0) \leqslant G(x) \leqslant \frac{1}{2k}\|x\|^2 + c$$

$$y^\star \in M \ : \quad G^\star(y^\star) < \frac{1}{2}\langle KSy^\star , y^\star\rangle$$

Then I has a non-zero pseudo-critical point $\bar{x} \in X$ ./

<u>Proof</u> : It has just been shown that has I has a pseudo-critical point $\bar{x} = KS\bar{x}^\star + \bar{y}$, where $\bar{y} \in \overline{K(\operatorname{Ker} A)}$ and $\bar{x}^\star$ minimizes $\bar{J}$ on M. If $\bar{x}$ were zero, we would get $KS\bar{x}^\star \in M^\perp$, and hence $\bar{J}(\bar{x}^\star) = G^\star(\bar{x}^\star)$. Since G

attains its minimum at the origin, so does $G^{\star}$. Since M contains both the origin and the minimizer $\bar{x}^{\star}$, the condition $\bar{x} = 0$ would imply that :

$$\bar{J}(\bar{x}^{\star}) = G^{\star}(0) = 0$$

and hence $\bar{J}(x^{\star}) \geqslant 0$ for all $x^{\star} \in M$. This means that

$$\forall x^{\star} \in M \quad , \quad G^{\star}(x^{\star}) \geqslant \frac{1}{2} \langle KSx^{\star}, x^{\star} \rangle$$

This contradicts our assumption on the existence of $y^{\star}$. Hence $\bar{x} \neq 0$, as desired ./

This certainly is a situation where duality is quite advantageous. Searching for a minimizer is computationally much simpler that searching for a saddle-point. Moreover, there may be more information to be gotten in specific situations from the fact that we are dealing with a minimizer : see [9] and [10] for instance.

Note for future reference that there is an a priori estimate on $\bar{x}^{\star}$, and hence on $\bar{v} = S\bar{x}^{\star}$. Indeed, we have :

$$\frac{1}{2} (k - \|KS\|) \|\bar{x}^{\star}\|^2 - c \leqslant J(\bar{x}^{\star}) \leqslant J(0) = G^{\star}(0)$$

and hence :

$$\|x^{\star}\|^2 \leqslant 2(G^{\star}(0) + c) (k - \|KS\|)^{-1}$$

B - SUPERQUADRATIC G

The situation in the superquadratic case is much more complicated. One of the main difficulties is that in the cases we are most interested in, K will be compact, so that the eigenvalues λ_n of KS converge to zero. It is therefore difficult to compare directly $\frac{1}{2} \langle KSx^{\star}, x^{\star} \rangle$ with $G^{\star}(x^{\star})$, even though the latter is subquadratic.

Moreover, whereas in the subquadratic case we are content with knowing the bahaviour of G at infinity, what little results we have concerning the superquadratic case require assumptions on the behavior of G near the origin. In other words, we have two types of existence theorems, under two types of assumptions on G :

- either subquadratic growth at infinity,
- or superquadratic behaviour near the origin.

We begin with the simple case when the dual functional can be shown to have a local minimum.

PROPOSITION 3 : Assumptions as in theorem 1-6. Assume moreover K is compact, G is differentiable at the origin, and there is a constant $k > 0$ such that, for all x in some neighborhood of zero :

$$G(x) \leqslant G(0) + \langle G'(0),x\rangle + \frac{1}{2k} \|x\|^2$$

Then there is some $\bar{\varepsilon} > 0$ such that, for $-\bar{\varepsilon} < \varepsilon < \bar{\varepsilon}$, the function :

$$I_\varepsilon(v) = \frac{1}{2} \langle Av,v\rangle + \varepsilon G(Kv)$$

has a pseudo-critical point of V ./

Note that the assumption on G is indeed satisfied (with any $k > 0$) if $G''(0)$ is well-defined and equal to zero. This is what we call superquadratic behaviour at zero.

Proof : The assumption on G implies a similar property on $G^\star$: there is a neighborhood $\mathcal{V}$ of $G'(0) = \bar{x}^\star$ such that, for all $x^\star$ in $\mathcal{V}$, we have :

$$G^\star(x^\star) \geqslant G^\star(\bar{x}^\star) + \frac{k}{2} \|x - \bar{x}^\star\|^2$$

We now apply theorem 1-6 to the function $I = \varepsilon^{-1} I_\varepsilon$ setting $A_\varepsilon = \varepsilon^{-1} A$ and hence $S_\varepsilon = \varepsilon S$. We obtain the dual function :

$$\bar{J}(x^\star) = -\frac{\varepsilon}{2} \langle KSx^\star,x^\star\rangle + G^\star(x^\star)$$

Setting $x^\star = \bar{x}^\star + y^\star$, with the new variable $y^\star$, we get :

$$\bar{J}(\bar{x}^\star+y^\star) \geqslant \frac{1}{2}(k - \varepsilon\|KS\|)\ \|y^\star\|^2 - \varepsilon\|KSx^\star\|\ \|y^\star\| + \bar{J}(\bar{x}^\star)$$

It is now clear that for $|\varepsilon|$ small enough, there will be some $\eta > 0$ such that :

$$\|y^\star\| = \eta \quad \Rightarrow \quad (\bar{x}^\star+y^\star) \in V \qquad \text{and} \qquad \bar{J}(\bar{x}^\star+y^\star) > \bar{J}(\bar{x}^\star)$$

Minimizing $\bar{J}(\bar{x}^\star+y^\star)$ for $\|y^\star\| \leqslant \eta$ then yields a local minimum for $\bar{J}$, and hence a weak critical point for I_ε . The fact that this minimum is attained follows from standard arguments, using the convexity of $G^\star$ and the compactness of K ./

The study of more general situations requires more sophisticated tools. Here is a typical result.

PROPOSITION 4 : Assumptions as in theorem 1-6. Assume moreover that K is compact and there is some $x_o^\star \in M$ where $\langle KSx_o^\star, x_o^\star\rangle > 0$, that G is strictly convex, attains its minimum at the origin with $G(0) = 0$, and satisfies :

$$\alpha^{-1} \sup\{G(x) \mid \|x\| \leqslant \alpha\} \to 0 \qquad \text{when } \alpha \to 0$$

Assume also there is some constant $k > 2$ such that :

$$\langle x, x^\star\rangle \geqslant k\, G(x) \qquad \forall\, x^\star \in \partial G(x)$$

Then $\bar{J}$ has a non-trivial critical point :

$$\exists\, \bar{x} \neq 0 \quad : \quad \bar{J}'(\bar{x}) = 0$$

Proof : By duality, the conditions on G yield conditions on $G^\star$, namely:

$$\inf\ \{G^\star(x^\star) \mid \|x^\star\| = \beta\} > 0 \qquad \text{for all } \beta > 0$$

$$\langle x, x^\star\rangle \leqslant k' G^\star(x^\star) \qquad \forall\, x \in \partial G^\star(x^\star)$$

Here k' is defined by $1/k + 1/k' = 1$, so that $1 < k' < 2$. This last condition can also be stated in another, equivalent way (see [11]):

$$\forall\, x^\star \in X^\star \;,\; \forall\, \lambda > 1 \;, \qquad G^\star(\lambda x^\star) \;\leqslant\; \lambda^{k'}\, G^\star(x^\star)$$

It follows easily that $\bar{J}(\gamma x_o) \to -\infty$ when $\gamma \to +\infty$. Moreover, choosing some $\beta_o > 0$ with

$$\delta_o \;=\; \inf\; \{G^\star(x^\star) \mid \|x^\star\| = \beta_o\} \;<\; +\infty$$

(such a β_o exists, otherwise G would be identically zero, and could not be strictly convex), we get :

$$\|x^\star\| \;\leqslant\; \beta_o \quad\Rightarrow\quad G^\star(x^\star) \;\geqslant\; \|x^\star\|^{k'}\, \beta_o^{-k'}\, \delta_o$$

It follows that we can find $\beta_1 > 0$ and $\gamma_o > 0$ such that :

(a) $$\inf\; \{\bar{J}(x^\star) \mid \|x^\star\| = \beta_1\} \;>\; 0 \;=\; \bar{J}(0)$$

(b) $$\bar{J}(\gamma_o x^\star) \;<\; 0 \;=\; \bar{J}(0)$$

It is a theorem of Ambrosetti and Rabinowitz ([2]) that if $\bar{J}$ satisfies these two conditions, and if in addition it is C^1 and satisfies condition (C) of Palais-Smale on M, then it has a non-trivial critical point :

$$\exists\; \bar{x}^\star \neq 0 \quad : \quad \bar{J}'(\bar{x}^\star) \;=\; 0$$

We refer to [20] for a proof that the same conclusion holds under the same conditions (a) and (b), provided only $\bar{J}$ is Gâteaux-differentiable, the derivative $\bar{J}'$ is continuous from V (strong) to $V^\star$ (weak) and the following condition is satisfied :

(weak C) $$\left\{ \begin{array}{l} \text{let } x_n^\star \text{ be a sequence in M such that } \bar{J}(x_n^\star) \text{ is bounded,} \\ \bar{J}'(x_n^\star) \neq 0 \text{ for all n , and } \bar{J}'(x_n^\star) \to 0 \text{ in } M^\star \text{ ; then there} \\ \text{is some point } \bar{x}^\star \in M \text{ where } J'(\bar{x}^\star) = 0 \text{ in } M^\star \text{ and} \\ \liminf\; \bar{J}(x_n^\star) \leqslant \bar{J}(x^\star) \leqslant \limsup\; \bar{J}(x_n^\star) \end{array} \right.$$

Here, since G is strictly convex, $\bar{J}$ is Gâteaux-differentiable and $\bar{J}'$ is strong-to-weak continuous. Now to check condition (weak C).

Take a sequence $x_n^\star \in M$ with

$$a \leqslant \bar{J}(x_n^\star) = -\frac{1}{2} <KSx_n^\star, x_n^\star> + G^\star(x_n^\star) \leqslant b$$

$$\bar{J}'(x_n^\star) = -KSx_n^\star + \partial G^\star(x_n^\star) = \xi_n \to 0 \quad \text{in } X$$

Substituting the second condition into the first :

$$a \leqslant \frac{1}{2} <\xi_n - \partial G^\star(x_n^\star) , x_n^\star> + G^\star(x_n^\star) \leqslant b$$

Using the growth condition on $G^\star$:

$$\left(1 - \frac{k'}{2}\right) G^\star(x_n^\star) + \frac{1}{2} <\xi_n, x_n^\star> \leqslant b$$

Since $k' < 2$ and $\xi_n \to 0$, the sequence $x_n^\star$ must be bounded. There is a subsequence, still denoted by $x_n^\star$, which converges weakly to some point $\bar{x}^\star$, which also belongs to M, since this is a closed subspace. Since K is compact, the sequence $KSx_n^\star$ converges strongly to $KS\bar{x}$, and $\partial G^\star(x_n^\star) = KSx_n^\star + \xi_n$ also converges to $KS\bar{x}$.

Fix $x^\star$ in M . We have, for each n :

$$G^\star(x^\star) \geqslant G^\star(x_n^\star) + <\partial G^\star(x_n^\star) , x^\star - x_n^\star>$$

Taking limits, we get

$$G^\star(x^\star) \geqslant G^\star(\bar{x}^\star) + <KS\bar{x}^\star, x^\star - \bar{x}^\star>$$

Hence $KS\bar{x}^\star = \partial G^\star(\bar{x}^\star) + M$, which means that $\bar{J}'(\bar{x}^\star) = 0$ in $M^\star$, as announced.

We have $G^\star(\bar{x}^\star) \leqslant \liminf G^\star(x_n^\star)$ by lower semi-continuity. We also have :

$$G^\star(\bar{x}^\star) \geqslant G^\star(x_n^\star) + <\partial G^\star(x_n^\star) , x^\star - x_n^\star>$$

so that $G^\star(\bar{x}^\star) \geqslant \limsup G^\star(x_n^\star)$. Finally $G^\star(\bar{x}) = \lim G^\star(x_n^\star)$ and condition (weak C) is established (remember that we have taken a subsequence of the original sequence) ./

Of course, under the assumptions of proposition 4, the origin also is a critical point (the trivial one) : $\bar{J}'(0) = 0$. The interest of proposition 4 lies in asserting the existence of a different one.

Theorem 1-6 associates with critical point of $\bar{J}$ pseudo-critical points of I. The trivial one, $x^\star = 0$, leads us to the equation $0 \in K^\star \partial G(0)$; in other words, $x = 0$ is trivially a pseudo-critical point of I. The other one, $\bar{x}^\star \neq 0$, leads us to a pseudo-critical point $\bar{x} = KS\bar{x}^\star + \bar{y}$; a special investigation, taking into account the particular situation, then is needed to decide whether $\bar{x} \neq 0$. For instance, as pointed out at the end of section I, if A(V) is closed in $V^\star$ and K(Ker A) closed in X, then $\bar{y} = K\bar{w}$ with $\bar{w} \in \text{Ker } A$, and $\bar{u} = S\bar{x}^\star + \bar{w}$ is a critical point of I. If $K^\star$ is injective on M, we get $AS\bar{x}^\star = - K^\star\bar{x}^\star \neq 0$, so that $A\bar{u} \neq 0$ and thus $\bar{u} \neq 0$.

III. THE WAVE EQUATION

We are given a number $T > 0$. We consider the space $\Omega = (0,1) \times R / TZ$, and we define :

$$V = \{u \in H^1(\Omega) \mid u(0,t) = 0 = u(1,t) \quad \forall\, t \in R/TZ\}$$

with the Hilbert structure induced by $H^1(\Omega)$. The functions $u(x,t)$ in V are T-periodic in t (time variable), and satisfy homogeneous Dirichlet conditions in x (space variable). They are supposed to represent the positions of a vibrating string of length unity, with fixed extremities.

All functions in $L^2(\Omega)$ have Fourier expansions

$$\phi = \Sigma \phi_{np} \sin(p\,\pi\, x) \exp\,(2\, i\, n\, \pi\, t/T)$$

$$p \geq 1\ ,\ n \in Z\ ,\quad \phi_{np} = \bar{\phi}_{(-n)p},\ \Sigma|\phi_{np}|^2 < \infty$$

The functions $u \in V$ are characterized by $\Sigma(1 + n^2 + p^2)\ |u_{np}|^2 < \infty$. We will study the wave operator $A : V \to V^\star$. It is defined by $Au = u_{tt} - u_{xx}$, or, in the Fourier expansion :

$$Au = \Sigma\ \pi^2(p^2 - 4\, n^2/T^2)u_{np} \sin(p\,\pi\, x) \exp\,(2\, i\, n\, \pi\, t/T)$$

Finally, we are given a real-valued function $g(u,x,t)$ on $R \times \Omega$. We assume that :

$$u \to g(u,x,t) \qquad \text{is convex}$$

$$(x,t) \to g(u,x,t) \qquad \text{is measurable}$$

We denote by $\partial g(u,x,t)$ the subgradient of g and by $g^{\star}(u,x,t)$ its Fenchel conjugate with respect to u, for fixed (x,t).

We begin with a short proof of a known result ([5], [6]).

PROPOSITION 1 : Assume T is rational, and g satisfies

$$b \in L^1 \; : \; g(u,x,t) \geqslant \frac{k_o}{2} |u|^2 - b(x,t)$$

$$c \in L^1 \; : \; g(u,x,t) \leqslant \frac{k_1}{2} |u|^2 + c(x,t)$$

Then there is some $\bar{u} \in L^2(\Omega)$ such that

$$(E) \qquad \bar{u}_{tt} - \bar{u}_{xx} + g(\bar{u},x,t) = 0$$

and $\bar{v} = \sum_{\left|\frac{n}{2}\right| \neq \frac{T}{2}} \bar{u}_{np} \sin(p\,\pi\,x) \exp\left(2\,i\,n\,\pi\,\frac{t}{T}\right)$ belongs to V ./

Proof : We wish to apply theorem 1-6. We first set up the framework of theorem 1-6. The space V and the operator A have just been defined ; it is easily checked that A is self-adjoint.

We define X to be $L^2(\Omega)$, so that it contains V, and K can be taken to be the standard injection of V into X. It is known to be compact. The range of the map $v \to (Av,Kv)$ over V is easily seen to be closed in $V^{\star} \times X$.

We now investigate the subspace $M = (K^{\star})^{-1} R(A)$ of $X^{\star}$, and try to define the map $S \in L(M,V)$. Remember that T is rational, $T = a/b$ for instance, with $a \in N$ and $b \in N$. The Fourier coefficient $(Au)_{np}$ of Au is given by :

$$(Au)_{np} = \pi^2 a^{-2} \left(a^2 p^2 - 4\,n^2 b^2\right) u_{np}$$

We thus have :

$$(Au)_{np} = 0 \quad \text{if} \quad 2|n|p^{-1} = ab^{-1} = T .$$

$$|(Au)_{np}| \leqslant \pi^2 a^{-2}(ap + 2|n|b)\ |u_{np}| \qquad \text{otherwise}$$

Now $\phi \in M$ if and only if $\phi \in L^2(\Omega)$ and $\phi = Au$ for some $u \in V$, which can be written :

$$\phi_{np} = (Au)_{np} \quad \text{with} \quad \Sigma(1+n^2+p^2)\ |u_{np}|^2 < \infty$$

We claim M is precisely the space of all $\phi \in L^2(\Omega)$ with $\phi_{np} = 0$ when $p = 0$ or $2|n|p^{-1} = T$. Indeed, the Fourier coefficients of u are then given, for $|n|p^{-1} \neq T$, by :

$$u_{np} = a^2\pi^{-2}(a^2p^2 - 4\ n^2b^2)^{-1}\ \phi_{np}$$

$$|u_{np}| \leqslant a^2\pi^{-2}\ (ap + 2|n|b)^{-1}\ |\phi_{np}|$$

Set $u_{np} = 0$ for $2|n|p^{-1} = T$. We have thus defined a function $u = S\phi$ by its Fourier coefficients. Since $\phi \in L^2(\Omega)$, we have $\Sigma|\phi_{np}|^2 < \infty$, and it easily follows that $\Sigma(n^2+p^2)|u_{np}|^2 < \infty$, so that $u \in V$.

We have thus shown that M is closed, and we have defined the map $S \in L(M,V)$. A more precise computation gives :

$$\|KS\| = \operatorname{Sup} \left\{\pi^{-2}\left(p^2 - \frac{4n^2}{T^2}\right)^{-1} \ \middle|\ 2|n|p^{-1} \neq T\right\} = \frac{T^2}{4\pi^2}$$

We now apply proposition 2-1, with :

$$G(u) = \int_\Omega g(u,(x,t),x,t)\ dx\ dt$$

Its conjugate $G^\star$ is known to be (see [15]) :

$$G^\star(u) = \int_\Omega g^\star(u(x,t),x,t)\ dx\ dt .$$

It follows from the assumptions that $\operatorname{Dom} G^\star = L^2(\Omega)$, which certainly contains the origin in its interior. Applying propositions 2-1, we find some $\psi \in M^\perp$ such that, setting $S\bar{x}^\star = \bar{v}$:

$$- A\bar{v} \in K^{\star}\partial G(K\bar{v}+\psi)$$

Since $\psi \in M^{\perp}$, we have $\psi_{np} = 0$ when $p \neq 2|n|T^{-1}$. It follows that $\psi_{tt} - \psi_{xx} = 0$ in the sense of distributions, and $K\bar{v} + \psi = \bar{u} \in L^2$ satisfies equation ($\mathcal{E}$). Note that the boundary conditions are lost ./

We now wish to get a similar result for irrational T. To do this, we introduce a continuous function $\alpha : R/2Z \to R$. We take it to be even, so that its Fourier expansion will be

$$\alpha(x) = \Sigma\alpha_p \cos (p \pi x) \quad , \quad p \geqslant 0$$

For any $\phi \in L^2(\Omega)$, we define $\tilde{\phi}$ to be the unique function on $R/2Z \times R/TZ$ such that $\tilde{\phi}(x,t) = \phi(x,t)$ for $0 < x < 1$ and $\tilde{\phi}(-x,t) = -\phi(x,t)$. In other words, $\tilde{\phi}$ has the same Fourier expansion as ϕ ,

$$\tilde{\phi}(x,t) = \Sigma\phi_{np} \sin (p \pi x) \exp (2 i n \pi x)$$

but it is valid over the torus $R/2Z \times R/TZ$. This enables us to define the convolute $\alpha \star \phi$:

$$(\alpha \star \phi)(x,t) = \int_{R/2Z} \alpha(y) \phi(x-y,t) dy$$

$$= \Sigma \frac{1}{2} \alpha_p \phi_{np} \sin(p \pi x) \exp (2 i n \pi t/T)$$

Note that $\alpha \star \phi$ is smoother than ϕ in general. It can be made as close to ϕ as one wants by choosing α close to the Dirac measure in $\mathcal{D}'(R/2Z)$.

Recall that the function g is measurable in (x,t) and convex in u. We will assume it to have subquadratic growth at infinity.

<u>PROPOSITION 2</u> : Assume g satisfies the following :

$$\forall \varepsilon > 0 , \exists \ell \in L^1(\Omega) : g(u,x,t) \leqslant \frac{\varepsilon}{2} |u|^2 + \ell(x,t)$$

$$\exists\, \phi^\star \in L^2(\Omega) \quad : \quad \int_\Omega g^\star(\phi^\star(x,t),x,)\, dt \; < +\infty$$

Assume $\Sigma|\alpha_p| < \infty$. Then, for almost every irrational $T > 0$, the equation

$$(E) \qquad u_{tt} - u_{xx} + \alpha \star \partial g(\alpha \star u,x,t) \ni 0$$

has a solution $u \in V$ ./

<u>Proof</u> : We again set up the framework of theorem 1-6. The space V and the operator A are as above.

We take $X = L^2(\Omega)$, and define $K : V \to X$ by $Ku = \alpha \star u$. It is clearly linear and compact. Its transpose $K^\star : X^\star \to V^\star$ is again the convolution by α , as defined above :

$$\forall\, \psi \in L^2(\Omega) \quad , \quad K^\star\psi = \alpha \star \psi$$

We now investigate the space M and try to define the map S. Since T is irrational, $(p^2 - 4\,n^2\,T^{-2})$ is never zero, so that A is one to one : Ker A = $\{0\}$. Moreover, it is well-known fact, the proof of which we postpone to a later lemma, that with almost every T one can associate a constant γ such that

$$\forall\, (q,p) \in Z \times N \quad , \quad |pT - q| \geqslant \gamma|\alpha_p|$$

We claim that $M = L^2(\Omega)$. To see this, take any $\psi \in L^2(\Omega)$ and try to solve the equation $K^\star\psi + Av = 0$. Equating Fourier coefficients, we get :

$$-\frac{1}{2}\,\alpha_p\,\psi_{np} = \pi^2(p^2 - 4\,n^2\,T^{-2})\; v_{np}$$

and hence :

$$v_{np} = -\frac{T^2}{2\pi^2}\;\frac{\alpha_p}{(p^2T^2 - 4\,n^2)}\;\psi_{np}$$

Writting $p^2T^2 - 4n^2 = (pT + |2|n)(pT - |2|n)$, and estimating the last term, we get :

$$|v_{np}| \leqslant \frac{T^2}{2\gamma\pi^2} \frac{1}{(pT + 2|n|)} |\psi_{np}|$$

It follows immediately that $n|v_{np}|$ and $p|v_{np}|$ are all bounded above by a $|\psi_{np}|$, where a is a constant depending only on T. Since $\psi \in L^2(\Omega)$, the sequence $|\psi_{np}|$ is quare-summable, and so are the sequences $n|v_{np}|$ and $p|v_{np}|$, so that v belongs to V. Setting $v = S\psi$, we have defined a map S with the desired properties, $\|S\|$ depending only on T.

We now define $G : X \to R$ by :

$$G(\phi) = \int_\Omega g(\phi(x,t),x,t)\, dx\, dt$$

Choosing $\varepsilon < \|KS\|^{-1}$, we have $G(\phi) \leqslant \frac{\varepsilon}{2} \|u\|^2 + \int_\Omega \ell\, dx\, dt$ and the result follows from proposition 2-1 ./

Note that, since $M^\perp = \{0\} = K$ (Ker A) , the critical points $\bar{v}$ of the dual functional $\bar{J}$ coïncide with the solutions $\bar{u}$ of the equation.

In the proof of proposition 2, we have used the following lemma, which is widely used for solving small divisors problems (see [21] for instance).

<u>Lemma</u> : Let α_p be a summable sequence. Then for almost every $t \in R$, there is some $\gamma > 0$ such that

$$\forall\ (q,p) \in Z \times N \quad , \quad |pt - q| \geqslant \gamma\, |\alpha_p|$$

<u>Proof</u> : For fixed $\gamma > 0$ and $a > 0$, we define

$$R(\gamma,a) = \{t \in]-a,a\,] \mid (q,p) \in Z \times N : |pt - q| < \gamma\, |\phi_p|\}$$

Clearly $R(\gamma,a) \subset R(\gamma',a)$ for $\gamma > \gamma'$. We wish to estimate the Lebesque measure of $R(\gamma,a)$: if we can prove that it goes to zero with γ, the result will follow immediately (first let $\gamma \to 0$, then $a \to \infty$).

We first fix p. There can be at most (2 ap + 3) intervals

$(qp^{-1} - \gamma|\alpha_p|p^{-1}, \ qp^{-1} + \gamma|\alpha_p|p^{-1})$ which meet the given interval $[-a,a]$, and their total length is $(2ap + 3)\ 2\gamma|\alpha_p|p^{-1}$.

Summing up over all $p \in N$, we get the desired estimate :

$$\mu\,[R(\gamma,a)] \leqslant \sum_p \frac{2ap + 3}{p}\ 2\gamma|\alpha_p| \leqslant 6(a+1)\gamma \sum_p |\alpha_p|$$

Since $\sum_p |\alpha_p|$ is finite, we see that $\mu\,[R(\gamma,a)]$ goes to zero with γ, as desired ./

Now assume $\partial g(0,x,t) \supset 0$ for all (x,t). Equation (E) then has the trivial solution $u = 0$, and we wish to find another one. Proposition 2-2 applies to give :

<u>PROPOSITION 3</u> : Assume moreover $g(0,x,t) = 0$ for all (x,t) and

$$\forall\, \varepsilon > 0 ,\ \exists\, \eta > 0 \ : \ |u| \leqslant \eta \ \Rightarrow\ g(u,x,t) \geqslant \frac{1}{2\varepsilon} u^2$$

If $\Sigma|\alpha_p| < \infty$, then for almost every $T > 0$, equation (E) has at least two solutions, 0 and $\bar{u}$ ./

<u>Proof</u> : We apply proposition 2-2. We have to check that there is some $\phi^\star \in L^2$ such that $\bar{J}(\phi^\star) < 0$. We have seen that :

$$\bar{J}(\phi^\star) = \int_0^T g^\star(\phi^\star(x,t),x,t)dx\,dt - \frac{T^2}{8\pi^2} \sum \frac{\alpha_p^2\,\psi_{np}^2}{(p^2T^2 - 4n^2)}$$

Choose some smooth $\phi_0^\star \in L^2$ such that the last term is strictly positive. Consider the function $\lambda \to \bar{J}(\lambda\phi_0^\star)$

$$\bar{J}(\lambda\phi_0^\star) = \int_0^T g^\star(\lambda\phi_0^\star(x,t),x,t)dx\,dt - \frac{\lambda^2}{2} \langle KS\ \phi_o^\star , \phi_o^\star \rangle$$

It follows from the assumption on $g^\star$ that $g^\star(v,x,t)v^{-2} \to 0$ when $v \to 0$, uniformly in (x,t). Since $\phi_0^\star$ is bounded, we can take $\lambda > 0$ so small that $\phi^\star = \lambda\phi_0^\star$ satisfies our claim ./

IV. DAMPED NONLINEAR OSCILLATIONS.

We want to study the following system of 2n ordinary differential equations :

$$(H) \qquad \begin{cases} \dfrac{dx_i}{dt} = \dfrac{\partial H}{\partial p_i}(t,x,p) \\[2ex] \dfrac{dp_i}{dt} = \dfrac{\partial H}{\partial x_i}(t,x,p) - ap_i \end{cases} \qquad \leqslant i \leqslant n$$

under various boundary conditions. Here $H = R \times R^n \times R^n \to R$ is a given function, called the Hamiltonian, and $a \in R$ is the damping coefficient. For $a = 0$ we get the usual Hamilton's equations. We assume $a \neq 0$ from now on.

In the case where $H(t,x,p) = \frac{1}{2} \sum p_i^2 + V(t,x)$, quite usual in classical mechanics, the equations (H) reduce to a second-order n-dimensional system,

$$\ddot{x} + a\,\dot{x} + \nabla V(t,x) = 0$$

Professor R. Broucke [7] has shown us that one can arrive at (H) simply by writing down the Euler-Lagrange extremality conditions for the integral

$$\int e^{at} \left(- \sum p_i \dot{x}_i + H(t,x,p) \right) dt$$

This will be the basis for the following analysis.

From now on, we simplify our notations by calling u the pair (x,p), an element of R^{2n}.

We begin by making standing assumptions on H :

$$(A)\quad \begin{cases} \forall\, t \in R \;,\; u \to H(t,u) \text{ is convex} \\ \forall\, u \in R \;,\; t \to H(t,u) \text{ is measurable} \\ \forall\, (c_1,c_2) \in R^{2n} \;,\; \int_0^T H^\star\left(t,c_1+c_2e^{-at}\right) e^{at}\, dt < +\infty \\ \exists\, u_0(t) : \int_0^T |u_0(t)|^2\, dt < \infty \text{ and } \int_0^T H(t,u_0(t))e^{at} dt < \infty \end{cases}$$

We denote by $\partial H(t,u)$ the subdifferential of $v \to H(t,v)$ at the point u , and by σ the linear operator in R^{2n} defined by $\sigma(x,p) = (p,-x)$. We rewrite the equations $(\mathcal{H})$ as follows :

$$(\mathcal{H}) \qquad \dot{u} \in \sigma\partial H(t,u) - (0,ap) \qquad \text{a.e.}$$

We are given a number $T > 0$, and we would like to find T-periodic solutions of system $(\mathcal{H})$. In view of Broucke's remark, ot seems natural to introduce the functional :

$$I(u) = -\int_0^T e^{at}\, p\, \dot{x}\, dt - \int_0^T e^{at}\, H(t,x,p)\, dt$$

over the space

$$V = \left\{u \;\middle|\; \int_0^T \dot{u}^2\, e^{at}\, dt < +\infty \;,\; \text{with } u(0) = u(T)\right\}$$

From now on, it is understood that the interval $[0,T]$ is endowed with the measure e^{at} dt , and L^2 will always refer to the space $L^2([0,T]\; ;\; e^{at}\, dt\; ;\; R^{2n})$. We identify V with a closed subspace of $L^2 \times L^2$ through the injection $i = u \to (\dot{u},u)$. Hence $V^\star$ is identified with the quotient of $L^2 \times L^2$ by the subspace $R(i)^\perp = N$:

$$N = \{(\phi,\psi) \in L^2 \times L^2 \mid \phi\dot{u} + \psi u = 0 \quad \forall\, u \in V\}$$

A few elementary calculations give the characterization :

$$(\phi,\psi) \in N \Rightarrow \dot{\phi} + a\phi = \psi \quad \text{and} \quad e^{aT}\phi(T) = \phi(0)$$

We now tackle the functional I on V. It does split into a quadratic form and a convex function. Our first task is to find the self-adjoint operator A which is to represent the quadratic form. We start from the obvious (non self-adjoint) representation :

$$Bu = [-2p,0,0,0] + N$$

$$\frac{1}{2}\langle Bu,u\rangle = \frac{1}{2}\int_0^T [-2p\dot{x} + 0\dot{p} + 0x + 0p]e^{at}\,dt =$$

$$= -\int_0^T e^{at}\,p\dot{x}\,dt$$

Clearly $B^{\star}u = [0,0,0,-2\dot{x}]$ and :

$$Au = \frac{1}{2}(B + B^{\star})u = [-p,0,0,-\dot{x}] + N$$

In this context, $K : V \to L^2$ is simply the second projection $(\dot{u},u) \to u$, so that $K^{\star} : L^2 \to V^{\star}$ will be the map $\phi \to (0,\phi) + N$.

Finally, we define $G : L^2 \to R \cup \{+\infty\}$ by :

$$G(\phi) = \int_0^T H(t,\phi(t))\,e^{at}\,dt$$

It follows from assumption A that it is convex, lower semicontinuous, and Dom $G^{\star} = L^2$

$$G^{\star}(\psi) = \int_0^T H^{\star}(t,\psi(t))\,e^{at}\,dt$$

$$\psi \in \partial G(\phi) \Rightarrow [\,(\psi,\phi) \in L^2 \times L^2 \text{ and } \psi(t) \in \partial H(t,\phi(t)) \text{ a.e. }]$$

We now test our construction : what is the meaning of the equation $I'(u) = 0$ for $u \in V$?

<u>Lemma 1</u>: The critical points of I are exactly the solutions $\bar{u}$ in V of the boundary-value problem

$$\dot{u} \in \partial H(t,u) - (0,ap)$$

$$x(0) = x(T)$$

$$p(0) = 0 = p(T) \quad ./$$

Proof : By definition, $\bar{u} \in B$ is a critical point of I if and only if

$$A\bar{u} + K^{\star}\partial G(K\bar{u}) \supset 0 \qquad \text{in } V^{\star}$$

this means that, setting $\bar{u} = (p,x)$

$$[-p,0,0,-\dot{x}\,] + [0,0,\psi_1,\psi_2] \in N$$

$$\psi = (\psi_1,\psi_2) \in L^2 \quad \text{and} \quad \psi(t) \in \partial H(t),\bar{u}(t)) \qquad \text{a.e.}$$

Using the characterization of N, this becomes :

$$-\dot{p} - ap = \psi_1$$

$$0 = -\dot{x} + \psi_2$$

$$-e^{at}\,p(T) = -p(0)$$

The first two equations yield system $(\mathcal{H})$. The last one, together with $\bar{u} \in V$, yields the boundary conditions ./

Note that we have 3 n boundary conditions for a 2n-dimensional system, so that it is very improbable that we can solve this problem ! Let us see, however, what our machinery tells us.

We first need to find M . We know that $\psi = (\psi_1,\psi_2)$ belongs to M if and only if we can solve the equation $K^{\star}\psi + Au = 0$ with u in V. This means :

$$[0,0,\psi_1,\psi_2] + [-p,0,0,-\dot{x}\,] \in N \quad , \quad u(0) = u(T)$$

which splits into :

$$-\dot{p} - ap = \psi_1 \quad , \quad e^{aT}p(T) = p(0) \quad , \quad p(T) = p(0)$$

$$0 = -x + \psi_2 \qquad , \qquad x(T) = x(0)$$

We easily derive the characterization :

$$M = \left\{ \psi \in L^2 \mid \int_0^T e^{at} \psi_1 \, dt = 0 \text{ and } \int_0^T \psi_2 \, dt = 0 \right\}$$

For all $\psi \in M$, we get :

$$x(t) = \int_0^t \psi_2(s) \, dt + \text{constant}$$

$$-p(t) = e^{-at} \int_0^t e^{as} \psi_1(s) \, ds \, .$$

There are many ways to choose the constant in the formula for x, the best way being the one which minimizes the norm of the corresponding linear operator $KS = L^2 \to L^2$. This done by setting $\int_0^T x \, dt = 0$, which gives :

$$x(t) = \int_0^t \psi_2(s) \, ds - \int_0^T dt \int_0^t \psi_2(s) \, ds$$

$$-p(t) = e^{-at} \int_0^t e^{as} \psi_1(s) \, ds \, .$$

This is the map $S : M \to V$; considering (x,p) as an element of L^2, we get the map $KS : M \to L^2$. We do not compute its norm ; we simply give it a name :

$$C(a,T) = \|KS\|$$

We are now ready to get pseudo-critical points. Let us first understang their meaning.

<u>Lemma 2</u> : Let $u \in L^2$ be a pseudo-critical point of I. Then $\dot{u} \in L^2$ and $u = (x,p)$ solves the boundary-value problem :

$$(P) \qquad \begin{cases} \dot{u} \in \sigma\partial H(t,u) - (0,ap) \\ x(0) = x(T) \\ p(0) = e^{aT} p(T) \end{cases}$$

Proof : By definition, we have $u = Kv + \psi$, with $Av + K^{\star}\partial G(u) \in 0$ and $\psi \in M$. The first condition has already been investigated in Lemma 1, where it was seen to mean, with $v = (y,q)$:

$$\dot{v} \in \sigma\partial H(t,u) - (0,aq)$$

$$y(0) = y(T)$$

$$q(0) = 0 = q(T)$$

The second condition, $\psi \in M$, obviously means that

$$\psi(t) = (c_1, c_2\, e^{-at}) \quad \text{with} \quad (c_1,c_2) \in R^{2n} .$$

Setting $u(t) = v(t) + \psi(t)$, we get the desired result. /

Our machinery has led us away from periodic solutions, towards another boundary-value problem : find motions wich return at the same position after time T, with momentum damped by a factor a. We immediately get existence results from section II :

Proposition 3 : Assumptions (A). Assume moreover that for some $k > C(a,T)$ and $c \in L^1$ we have :

$$H(t,u) \leqslant \frac{1}{2k} |u|^2 + c(t)$$

Then problem ($\mathcal{P}$) has a solution ./

Proof : Just apply proposition 2-1. The only non-trivial condition to check is $0 \in \text{Int}(\text{Dom}\, G^{\star} - M)$. Byt assumption (A) tells us in effect that $\text{Dom}\, G^{\star} \subset M$, so $\text{Dom}\, G^{\star} - M$ is the whole of L^2. Hence the result ./

Corollary : Assumptions (A). Assume moreover that there is a function $\phi : R_+ \to R_+$, with $\phi(u)u^{-2} \to 0$ when $u \to +\infty$, and a constant $c \in R$ such that :

$$H(t,u) \leqslant \phi(|u|) + c$$

Then problem $(\mathcal{P})$ has a solution for all $T > 0$ ./

Proposition 4 : Assumptions (A). Assume that H does not depend on t, and that $H(u) \geqslant H(0) = 0$ for all $u \in R^{2n}$. Assume moreover that we have :

$$\liminf_{u \to 0} H(u)u^{-2} = +\infty$$

$$\limsup_{u \to \infty} H(u)u^{-2} = 0$$

Then the problem $(\mathcal{P})$ has a non-trivial solution for all $T > 0$ ./

Proof : Just apply proposition 2-2. All we have to do is to check the conditions G. The first one is trivial. For the second, we know that for all $c > 0$ there is some $\varepsilon > 0$ such that

$$|u| \leqslant \varepsilon \Rightarrow H(u) \geqslant \frac{c}{2} |u|^2 .$$

It follows that there is some $\eta > 0$ such that :

$$|v| \leqslant \eta \Rightarrow H^{\star}(v) \leqslant \frac{1}{2c} |v|^2 .$$

One easily finds a smooth function w with $\langle KSw,w\rangle > 0$.

If we choose $c^{-1} < \langle KSw,w\rangle$, the function $t \to \bar{J}(tw)$ will assume negative values for small t. Hence the result ./

We conclude with the superquadratic case :

Proposition 5 : Assume that $H : R^{2N} \to R$ is strictly convex, that $H(u) \geqslant H(0) = 0$ for all $u \in R^{2n}$, and that, for some $k > 2$, we have :

$$\forall \lambda > 1 \qquad H(\lambda u) \geqslant \lambda^k H(u)$$

Then the problem $(\mathcal{P})$ has a non-trivial solution for all $T > 0$.

Proof : Assume first that H satisfies a supplementary condition :

$$\exists\, a > 0 \ : \ H(u) \leqslant a|u|^k$$

Take $V = \{u \mid \dot{u} \in L^k \ , u(0) = u(T)\}$ and $X = L^k$ (always with the measure $e^{at}dt$). Define A and G as above. By proposition 2-4, the functional $\overline{J}$ will have a non-zero critical point $\phi^\star \in L^{k'}$. Then there is some $\psi \in M$ such that $KS\phi^\star + \psi = u$ is a pseudo-critical point for I. Setting $u = (x,p)$, and recalling what S and M are in this case, we see that x cannot be constant. Hence the result.

BIBLIOGRAPHIE

[1] AMBROSETTI - MANCINI. "Solutions of minimal period for a class of convex Hamiltonian systems". J. Diff. Eq., 1982.

[2] AMBROSETTI - RABINOWITZ. "Dual variational methods in critical point theory and applications". J. Func. An. 14 (1973) 249-381.

[3] AUBIN. "Mathematical methods of game and economic theory". North-Holland - Elsevier 1980.

[4] AUBIN - EKELAND. "Second-order evolution equations associated with convex Hamiltonians", Can. Math. Bull. 23 (1980), 81-94.

[5] BREZIS - CORON - NIRENBERG. "Free vibrations for a nonlinear wave equation and a theorem of P. Rabinowitz", Comm. Pure Appl. Math, 33 (1980), 667-684.

[6] BREZIS - CORON. "Periodic solutions of nonlinear wave equations and Hamiltonian systems", Ann. J. of Math., 103 (1980), 559-570.

[7] BROUCKE. Personal communication, 1979.

[8] CLARKE. "Periodic solutions to Hamiltonian inclusions", J. Diff. Eq., 34 (1979), p. 523-534.

[9] CLARKE - EKELAND. "Hamiltonian trajectories having prescribed minimal period", Comm. Pure Appl. Math., 33 (1980), 103-116.

[10] CLARKE - EKELAND. "Nonlinear oscillations and boundary-value problems for Hamiltonian systems", J. Rat. Mech. An., 1982.

[11] EKELAND. "Periodic solutions of Hamiltonian equations and a theorem of P. Rabinowitz", J. Diff. Eq. 34 (1979), p. 523-534.

[12] EKELAND. "Forced oscillations of nonlinear Hamiltonian systems II", Advances in Math., Nachbin ed., 1981, Academic Press.

[13] EKELAND. "Oscillations de systèmes hamiltoniens non linéaires III", Bulletin Soc. Math. de France, 1982.

[14] EKELAND - LASRY. "On the number of periodic trajectories for a Hamiltonian flow on a convex energy surface", Annals of Math., 112, 1980, p. 283-319.

[15] EKELAND - TEMAM. "Convex analysis and variational problems", North-Holland - Elsevier, 1976.

[16] ROCKAFELLAR. "Convex analysis", Princeton University Press, 1970.

[17] TOLAND. "A duality principle for non-convex optimisation and the calculus of variations", Arch. Rat. Mech. An.

[18] TOLAND. "Duality in non-convex optimisation", Journal Math. An. Appl..

[19] WILLEM. "Remarks on the dual leastaction principle", To appear, Zeitschrif für An. und ihre Anw..

[20] AUBIN - EKELAND. "Nonlinear analysis", Wiley, to appear.

[21] MOSER - SIEGEL. "Lectures on clestial mechanics", Springer 1971.

LA THEORIE DE LA SECONDE VARIATION ET LE PROBLEME LINEAIRE QUADRATIQUE

Pierre BERNHARD

Résumé

Dans le cadre de la commande optimale, on donne la démonstration de quelques faits classiques du calcul des variations, en les étendant au cadre présent. On étudie notamment le rôle des points conjugués, et on améliore le traitement du cas anormal en le rattachant à la commandabilité.

Introduction

Cette note est pour l'essentiel une mise en forme, dans le langage de la commande optimale, de résultats présents, de manière plus ou moins explicite, dans C. CARATHEODORY [4]. Par rapport à cette référence, on a étendu le traitement $x(t_1) = 0$ à $x(t_1) \in M$, sous espace vectoriel, et on a amélioré le traitement du cas anormal, Carathéodory ne disposant pas des outils de la théorie des systèmes nécessaires pour élucider complètement ce cas. Les résultats présentés sont largement réminiscents d'un cours professé par R.E. KALMAN en 1968, mais ne semblent pas disponibles sous forme imprimée à ce jour.

Dans la première partie, on rappelle des résultats considérés comme élémentaires sur le "problème auxiliaire" et le problème linéaire quadratique. A ce propos, on renvoie à [1] pour plus de détails. Dans une deuxième partie, on règle la question de la normalité des trajectoires, dans la troisième partie on introduit l'intégrale complète du problème linéaire-quadratique, qu'on explicite beaucoup plus que Carathéodory, pour établir l'optimalité du champ jusqu'à un point focal. La théorie de l'intégrale complète présente l'intérêt de pousser plus loin le parallèle entre l'équation de RICCATI et la théorie associée d'une part, et les équations canoniques de l'autre. Enfin, on établit les deux résultats fondamentaux du Calcul des Variations quant aux points conjugués, ce dans un contexte plus large que dans la théorie classique.

Nous utilisons la notation "$\cdot$" pour $\frac{"d"}{dt}$, et "'" pour "transposé". Un élément de R^n est considéré comme un vecteur colonne, et pour $f: R^n \to R$, $\frac{\partial f}{\partial x}$ est un vecteur ligne.

1. Le problème auxiliaire

1.1. La deuxième variation et le problème auxiliaire

1.1.1 Conditions nécessaires d'optimalité

On considère un problème de commande optimale dans R^n, donné par le système.

$$\dot{x} = f(x,u,t) \qquad x(t_0) = x_0 \qquad \phi(x(t_1)) = 0 \tag{1}$$

où f est C^2 en (x,u,t). t_1 est un instant final fixe, ϕ est C^2 de R^n dans R^1. Les commandes admissibles sont les fonctions continues par morceaux.

$$u(.) : t \to u(t) \in U \subset R^m. \quad \text{convexe fermé} \tag{2}$$

On pourrait, sans difficulté, étendre au cas où U dépend de T, mais pas de x. On veut minimiser le critère

$$J(x_0,t_0; u(.)) = K(x(t_1)) + \int_{t_0}^{t_1} L(x(t),u(t),t)\, dt. \tag{3}$$

K et L sont des fonctions réelles de classe C^2.

A ce problème, on associe le système adjoint, portant sur une fonction vectorielle de $\mathbb{R}$ dans $\mathbb{R}^n$, $\lambda(t)$:

$$\dot{\lambda}' = -\lambda'\frac{\partial f}{\partial x} - \lambda_0 \frac{\partial L}{\partial x}, \qquad \lambda'(t_1) = \lambda_0 \frac{\partial K}{\partial x} + \nu' \frac{\partial \phi}{\partial x} \tag{4}$$

(λ_0 est une constante qu'on prend égale à 1 chaque fois qu'elle n'est pas nulle) et on définit l'hamiltonien.

$$H(x,\lambda,u,t) = \lambda_0 L(x,u,t) + \lambda' f(x,u,t). \tag{5}$$

On rappelle le théorème ci-dessous (preuve dans [6]):

PRINCIPE DU MINIMUM (PONTRYAGUINE) Une condition nécessaire pour qu'une commande u(.) admissible soit optimale est qu'il existe $\lambda_0 \geq 0$ et $\lambda(t)$ continu satisfaisant (4) tel que le long de la trajectoire engendrée par u(.), on ait

$$H(x(t),\lambda(t),u(t),t) \leq H(x(t),\lambda(t),v,t) \quad \forall t,\ \forall v \in U. \tag{6}$$

La forme faible de cette condition, qui est une conséquence de (6), est

$$\frac{\partial H}{\partial u}\,\delta u \geq 0 , \qquad \forall \delta u : u + \delta u \in U . \tag{7}$$

En effet, U est supposé convexe.

1.1.2 Variations de J

Nous évaluons maintenant les deux premiers termes du développement de Taylor de J autour d'une valeur nominale J (u(.)) :

$$\delta J = \frac{dJ}{du} \, . \, \delta u(.) + \frac{1}{2} \left(\frac{d^2J}{du^2} \, . \, \delta u(.)\right) \, . \, \delta u(.)$$

pour une "variation faible" $\delta u(.)$ donnée, qu'on fera ensuite tendre vers zéro dans L^2.

On sait, par application des théorèmes classiques d'analyse, que δJ peut se calculer en développant au second ordre les équations (1) et (3). (Le terme x(.) qu'on calcule ainsi étant lui aussi la somme des deux premiers termes du développement de Taylor de x(.) autour de sa valeur nominale).

$$\delta \dot{x}_i = \frac{\partial f_i}{\partial x} \delta x + \frac{\partial f_i}{\partial u} \delta u + \frac{1}{2} \delta x' \frac{\partial^2 f_i}{\partial x^2} \delta x + \delta u' \frac{\partial^2 f_i}{\partial u \partial x} \delta x + \frac{1}{2} \delta u' \frac{\partial^2 f_i}{\partial u^2} \delta u \qquad (8)$$

$$\delta J = \frac{dK}{dx} \delta x_1 + \int_{t_0}^{t_1} \frac{\partial L}{\partial x} \delta x + \frac{\partial L}{\partial u} \delta u dt + \frac{1}{2} \delta x'_1 \frac{\partial^2 K}{\partial x_1^2} \delta x_1 + \frac{1}{2} \int_{t_0}^{t_1} (\delta x' \delta u') \begin{pmatrix} \frac{\partial^2 L}{\partial x^2} & \frac{\partial^2 L}{\partial x \partial u} \\ \frac{\partial^2 L}{\partial u \partial x} & \frac{\partial^2 L}{\partial u^2} \end{pmatrix} \begin{pmatrix} \delta x \\ \delta u \end{pmatrix} dt \qquad (9)$$

En introduisant λ défini par (4), et en remarquant que

$$\frac{d\phi}{dx_1} \delta x_1 + \frac{1}{2} \delta x'_1 \frac{d^2\phi}{dx_1^2} \delta x_1 = 0 \, (\| \delta u \|^3)$$

on peut reécrire (9) sous la forme

$$\delta J = \int_{t_0}^{t_1} \frac{\partial H}{\partial u} \delta u \, dt + \frac{1}{2} \delta x'_1 \frac{d^2(K + \nu'\phi)}{dx_1^2} \delta x_1 + \frac{1}{2} \int_{t_0}^{t_1} (\delta x' \delta u') \begin{pmatrix} \frac{\partial^2 H}{\partial x^2} & \frac{\partial^2 H}{\partial x \partial u} \\ \frac{\partial^2 H}{\partial u \partial x} & \frac{\partial^2 H}{\partial u^2} \end{pmatrix} \begin{pmatrix} \delta x \\ \delta u \end{pmatrix} \qquad (10)$$

Une conséquence du principe du minimum est que, si u(.) est optimal, et <u>intérieur à U pour tout t</u>, alors $\partial H / \partial u$ est nul.

Dans le cas où U est fermé, et défini par

$g_i(u) \leq 0, \quad i = 1, \ldots, p$

on se restreint, pour l'étude de l'optimalité locale de u (.), à des variations de la forme $\varepsilon\delta u(.)$, avec

$$\frac{\partial H}{\partial u} \delta u(t) = 0 \text{ presque tout } t,$$

car sinon, d'après (7) et (10) on a $\delta J > 0$.

Les g_i étant supposés dérivables, et de jacobien partout de plein rang, on sait qu'il existe des multiplicateurs $\mu_i(t)$ tels que

$$\frac{\partial H}{\partial u}(t) + \Sigma\, \mu \frac{\partial g_i}{\partial u}(t) = 0$$

les μ_i étant nuls pour les g_i négatifs, et positifs ou nuls pour les autres. On va donc se restreindre à des directions $\delta u(.)$ satisfaisant.

$$\frac{\partial g_i}{\partial u} \delta u(t) = 0 \qquad \forall i : g_i(u(t)) = 0.$$

Ainsi, si l'ensemble des contraintes saturées est constant par morceaux, tout se passe comme si u(t) était intérieur à U, à condition de restreindre les variations considérées au sous espace ainsi défini. Donc toute la théorie qui suit s'applique, à condition de restreindre à chaque instant les opérateurs G, R et S ci-dessous au sous espace tangent aux contraintes saturées.

Donc la condition "si u(t) est intérieur à U" pourra toujours être interprétée ainsi.

On va alors étudier les termes de second ordre dans (10). Remarquons alors qu'on peut, pour les évaluer, ne calculer δx qu'au premier ordre. Introduisons alors les notations (où G, S, R sont restreints au sous espace de R^m tangent à U en u au sens précis indiqué ci-dessus).

$$\frac{\partial f}{\partial x} = F\,, \quad \frac{\partial f}{\partial u} = G\,, \quad \frac{\partial \phi}{\partial x} = \Phi$$

$$\frac{\partial^2 H}{\partial x^2} = Q\,, \quad \frac{\partial}{\partial u}\left(\frac{\partial H}{\partial x}\right)' = S\,, \quad \frac{\partial^2 H}{\partial u^2} = R\,, \quad \frac{d^2(K + \nu'\phi)}{\partial x^2} = A$$

et le problème auxiliaire :

$$\dot{x} = Fx + Gu \quad x(t_0) = x_0 \quad \Phi x(t_0) = 0 \tag{14}$$

$$2J = x'(t_1)\ Ax(t_1) + \int_{t_0}^{t_1} (x'u') \begin{pmatrix} Q & S \\ S' & R \end{pmatrix} \begin{pmatrix} x \\ u \end{pmatrix} dt \tag{15}$$

on a le résultat suivant :

THEOREME 1.1. La trajectoire considérée est un optimum local pour des variations faibles si et seulement si le problème auxiliaire associé, initialisé en $x_0 = 0$ admet une commande optimale.

DEMONSTRATION Le problème auxiliaire est identique, aux notations près, à (8) limité au premier ordre et (10) (11). On montre ci-dessous que s'il a une commande optimale, c'est nécessairement $u(t) = 0$. Comme elle conduit à $J = 0$, on en déduit dans (8) (10) que pour $\delta u(.) \neq 0$, $\delta J \geq 0$. Réciproquement, si $u = 0$ n'est pas optimale, dans (14) (15), il existe dans (8)(10) des $u(.)$ tels que $\delta J \leq 0$, contredisant le fait que la trajectoire considérée soit optimale.

1.2. La théorie de Carathéodory et le problème auxiliaire

1.2.1. Conditions suffisantes d'optimalité

Au problème posé en 1.1.1, on associe de la façon suivante l'équation d'Hamilton-Jacobi-(Carathéodory-Isaacs-Bellman). Soit

$$\overline{H}(x,\lambda,t) = \min_{u \in U} H(x,\lambda,u,t) \tag{16}$$

et $\overline{u}(x,\lambda,t)$ un argument de ce minimum. (On rappelle que U est fermé). On introduit l'équation annoncée qui est une équation aux dérivées partielles portant sur une fonction réelle $V(x,t)$

$$\frac{\partial V}{\partial t} + \overline{H}(x, \frac{\partial V}{\partial x}, t) = 0 \quad , \quad V(x,t_1)\Big|_C = K(x) \ . \tag{17}$$

Ici, C est la cible : $C = \{x | \Phi(x)\} = 0$.

A toute fonction V , (16) fait correspondre une stratégie :

$$u^*(x,t) = \overline{u}(x, \frac{\partial V}{\partial x}(x,t), (t) \tag{18}$$

et on notera souvent par une étoile les quantités où u est remplacé par u^*, et λ par $\frac{\partial V}{\partial x}$. On dira que u^* transfère (x_0,t_0) sur la cible si l'équation (1) où on a remplacé u par u^* admet une solution unique telle que

$\phi(x(t_1)) = 0$. On appellera Ω un voisinage de la trajectoire correspondante dans R^{n+1}.

On rappelle alors le théorème suivant :

THEOREME DE CARATHEODORY <u>S'il existe une fonction $V(x,t)$ de classe C^1, telle que la stratégie correspondante transfère (x_0,t_0) sur la cible, et qui satisfait</u> (17) <u>dans un voisinage</u> Ω <u>de la trajectoire correspondante</u> $x^*(t)$, <u>alors la commande</u> $u^*(x^*(t),t)$ <u>est optimale parmi toutes les commandes admissibles pour lesquelles les trajectoires restent dans</u> Ω. <u>La valeur optimale du critère est</u> $J^* = V(x_0,t_0)$.

DEMONSTRATION Ici la preuve est tellement simple que nous la donnons. Remarquons que

$$\frac{\partial V}{\partial t} + H(x, \frac{\partial V}{\partial x}, u, t) = \frac{dV}{dt} + L$$

où $\frac{dV}{dt}$ est la dérivée de la fonction $t \to V(x(t),t)$ évaluée le long de la trajectoire engendrée par u, x(t) étant solution de (1). D'après (16) et (17), on a

$$\frac{dV}{dt} + L \geq 0 , \qquad \forall u(.)$$

et en intégrant de t_0 à t_1, et en tenant compte de la condition finale

$$J(x_0,t_0; u(.)) \geq V(x_0,t_0) \quad ,$$

pour tout u(.) transférant l'état sur la cible. D'autre part, pour $u^0(t) = u^*(x^*(t),t)$, on a l'égalité dans ces deux dernières relations.

REMARQUE 1. On a énoncé une version locale du théorème, mais en prenant $\Omega = R^n \times [t_0,t_1]$ on a une version globale.

REMARQUE 2. Si l'énoncé est local en x, il est global en u : nous permettons des <u>variations fortes</u> (cf. [1]).

1.2.2. Les équations canoniques

On cherche à résoudre (17) par la théorie des caractéristiques. Les équations des caractéristiques de (17) sont précisément (1) et (4), avec $\partial V/\partial x = \lambda$. Nous considérons ces équations munies de leurs <u>condi-</u>

tions finales. Ces conditions représentent une variété de dimension n dans R^{2n}. Soit

$$x = \xi_1(\rho) \qquad \rho \in D \subset \mathbb{R}^n \qquad \lambda = \ell_1(\rho)$$

un paramétrage. Nous considérons la famille de trajectoires :

$$x(t) = \xi(t,\rho), \qquad \xi(t_1,\ell) = \xi_1(\rho)$$

$$\lambda(t) = \ell(t,\rho) \qquad \ell(t_1,\rho) = \ell_1(\rho)$$

engendrées par les équations différentielles (1) et (4) depuis ces conditions finales, et

$$v(t,\rho) = \int_t^{t_1} L(\xi(t,\rho), \bar{u}(\xi(t,\rho), \ell(t,\rho), t), t)\, dt + K(\xi(t_1,\rho)).$$

engendrées par les équations différentielles (1) et (4) depuis ces conditions finales, et

$$v(t,\rho) = \int_t^{t_1} L(\xi(t,\rho), \bar{u}(\xi(t,\rho), \ell(t,\rho), t), t)\, dt + K(\xi(t_1,\rho)).$$

Si

$$X(t) = \frac{\partial\xi(t,\rho)}{\partial\rho} \tag{19}$$

est inversible pour ρ_0, sur $[t_0,t_1]$, localement on peut inverser

$$x(t) = \xi(t,\rho) \Rightarrow \rho = r(t,x(t))$$

et donc dans un voisinage Ω de la trajectoire $\xi(t,\rho_0)$, v définit une fonction

$$V(x,t) = v(t,r(t,x))$$

Une condition nécessaire pour que V soit, localement, solution de (17) est que $\lambda = \ell(t,\rho)$ soit bien son gradient, ce qui suppose, pour tout t

$$\lambda' = \frac{\partial V}{\partial x} = \frac{\partial v}{\partial \rho}\left(\frac{\partial\xi}{\partial\rho}\right)^{-1}$$

D'où

$$\lambda'\frac{\partial\xi}{\partial\rho_i} = \frac{\partial v}{\partial\rho_i}$$

En dérivant une deuxième fois en ρ, il vient

$$\frac{\partial \ell'}{\partial \rho_i} \frac{\partial \xi}{\partial \rho_j} = \frac{\partial^2 v}{\partial \rho_i \partial \rho_j} - \ell' \frac{\partial^2 \xi}{\partial \rho_i \partial \rho_j}$$

Le second membre est symétrique en i et j, le premier doit donc l'être aussi. Si on pose

$$\Lambda(t) = \frac{\partial \ell(t)}{\partial \rho}$$

on obtient la condition nécessaire

$$\Lambda' X = X' \Lambda$$

On démontre (cf. [4]) que la condition (21) qui exprime que les crochets de Lagrange de la paire (ξ, ℓ) :

$$[i,j] = \sum_k \frac{\partial \xi_k}{\partial \rho_i} \frac{\partial \ell_k}{\partial \rho_j} - \frac{\partial \xi_k}{\partial \rho_j} \frac{\partial \ell_k}{\partial \rho_i}$$

sont nuls, est aussi suffisante pour que V soit solution de l'équation (17). On peut alors appliquer le théorème de Carathéodory dans ce voisinage.

Les deux conditions à remplir sont donc (21), dont on va montrer qu'il suffit de la vérifier en t_1, et

$$\det X(t) \neq 0 , \quad \forall t \in [t_0, t_1] \ldots \tag{22}$$

Cette deuxième condition est beaucoup plus difficile à étudier, et fait l'objet de ce qui suit. On voit tout de suite qu'elle ne peut être satisfaite en t_1 que si $\phi \equiv 0$. Mais nous relaxerons la condition (22) en éliminant t_1 dans la partie 3.

En appliquant la théorie classique des équations différentielles, on établit facilement les équations différentielles satisfaites par X et Λ. Supposons $\bar{u}$ unique et dérivable en x et en λ :

$$\dot{X} = \frac{\partial X}{\partial t} = \frac{\partial}{\partial t} \frac{\partial \xi}{\partial \rho} = \frac{\partial}{\partial \rho} \frac{\partial \xi}{\partial t} = \frac{\partial}{\partial \rho} (f(\xi, \bar{u}(\xi, \ell, t), t)$$

$$= \frac{\partial f}{\partial x} X + \frac{\partial f}{\partial u} \frac{\partial \bar{u}}{\partial x} X + \frac{\partial \bar{u}}{\partial \lambda} \Lambda$$

et de même pour $\dot{\Lambda}$. En utilisant $\partial H/\partial u = 0$ qui découle de (6) et du développement de la partie (1.1.2.), pour évaluer les dérivées partielles de $\bar{u}$, on arrive (avec les notations (12) (13)) aux équations ci-dessous, appelées <u>équations canoniques</u> :

$$\dot{X} = (F - GR^{-1}S')X - GR^{-1}G'\Lambda \tag{23}$$

$$\dot{\Lambda} = -(Q-SR^{-1}S')X - (F'-SR^{-1}G')\Lambda \tag{24}$$

et les conditions finales donnent

$$\Phi X(t_1) = 0 \qquad \Lambda(t_1) = AX(t_1) + \Phi'N \tag{25}$$

et on doit en outre satisfaire (21) en $t = t_1$

On a en effet le résultat suivant :

LEMME 1.1 Si $\Lambda'(t\ )\ X(t\ )$ <u>est symétrique, les équations différentielles</u> (23) (24) <u>engendrent des matrices</u> $\Lambda(t)$ <u>et</u> $X(t)$ <u>telles que</u> $\Lambda'(t)\ X(t)$ <u>reste symétrique</u>.

PREUVE Il suffit d'utiliser (23) et (24) pour calculer

$$(\Lambda'X)^{\cdot} = -X'(Q-SR^{-1}S')X - \Lambda'GR^{-1}G'\Lambda \tag{26}$$

qui montre le résultat, car Q et R sont des matrices symétriques.

On a le résultat suivant :

THEOREME 1.2 <u>Si le problème auxiliaire</u> (14)(15) <u>a une solution, la trajectoire correspondante est localement optimale vis-à-vis des variations fortes</u>.

DEMONSTRATION On verra ci-dessous que le fait que le problème auxiliaire ait une solution est équivalent au fait qu'il existe une solution de (23) (24) satisfaisant (21) et (22) et (25). Nous étendrons le résultat à $X(t_1)$ singulière dans la partie 3.

REMARQUE importante. Les calculs faits ci-dessus supposent R inversible. La condition (6) implique $R \geq 0$ (condition de Legendre). Nous nous restreignons dans toute cette étude à $R > 0$. Nous renvoyons le lecteur à [1] pour voir comment et dans quelle mesure le cas singulier $\det R = 0$

se ramène au cas régulier.

En conclusion de ces deux premières sections, nous avons jusqu'ici précisé dans quelle mesure l'étude du problème auxiliaire permet de conclure à l'optimalité de la trajectoire considérée. Toute la suite est consacrée à ce problème auxiliaire ou problème linéaire quadratique.

1.3 Théorie élémentaire du problème linéaire quadratique. (cf.[5])

1.3.1 Equation de Riccati

Nous considérons ici le problème quadratique à état final libre :

$$\dot{x} = Fx + Gu \qquad x(t_0) = x_0 \tag{27}$$

$$2J = x'(t_1)\,Ax(t_1) + \int_{t_0}^{t_1} (x'u') \begin{pmatrix} Q & S \\ S' & R \end{pmatrix} \begin{pmatrix} x \\ u \end{pmatrix} dt \quad , \quad R > 0 \tag{28}$$

Les matrices F, G, Q, R et S peuvent dépendre de temps, continûment par morceaux. Nous écrivons l'équation d'Hamilton-Jacobi-etc. et plaçons arbitrairement

$$V(x,t) = \frac{1}{2}\, x'P(t)x \ ; \qquad P \text{ matrice carrée symétrique} \tag{29}$$

d'où $\lambda' = \frac{\partial V}{\partial x} = Px.$

L'équation (17) dégénère alors en une équation différentielle sur P, dire équation de Riccati :

$$\dot{P} + PF + F'P - (PG+S)\,R^{-1}(G'P+S') + Q = 0 \ , \ P(t_1) = A \tag{30}$$

de sorte que le théorème de Carathéodory s'énonce ici :

THEOREME 1.3 *Si l'équation de Riccati* (30) *admet une solution sur* t_0, t_1 , *alors le problème* (27)(28) *admet une solution optimale donnée par*

$$u^*(x,t) = -R^{-1}(G'P+S')x \tag{31}$$

$$J(x_0,t_0;u^*) = \frac{1}{2}\, x_0'P(t_0)x_0$$

DEMONSTRATION Il reste juste à vérifier que l'expression (31) est bien celle qui minimise H, ce qui estlaissé au lecteur. L'autre expression est juste (29) utilisée en t_0.

Il n'est pas inintéressant pour la suite de généraliser très légèrement au cas où le terme final du critère comporte une partie linéaire:

$$J = \frac{1}{2} x'(t_1) Ax(t_1) + a'x(t_1) + \frac{1}{2}\int_{t_0}^{t_1} (x'u') \begin{pmatrix} Q & S \\ S' & R \end{pmatrix} \begin{pmatrix} x \\ u \end{pmatrix} dt \qquad (32)$$

On vérifie alors, en posant

$$2V(X,t) = x'P(t) x + 2g'(t) x + h(t) \quad , \quad u^* = - R^{-1} G'(Px+g) \qquad (33)$$

qu'on a une solution de l'équation (17) si P satisfaisait encore (30), et

$$\dot{g} + (F' - SR^{-1} G')g = 0 \quad , \quad g(t\ ) = a \qquad (34)$$

$$\dot{h} = g'GR^{-1} G'g \quad , \quad h(t_1) = 0 \qquad (35)$$

Pour la prise en compte d'autres termes non homogènes dans les équations (excitation connue dans le dynamique et termes linéaires dans la partie intégrale du critère) on n'a juste qu'à compléter (34) et (35). Comme on n'en aura pas besoin ici, on renvoie le lecteur intéressé à [1].

1.3.2. Les équations canoniques

On considère le même problème, mais on autorise en outre une contrainte du type

$$\Phi x(t_1) = 0 \quad , \quad \Phi \text{ matrice } \ell \times n \text{ de plein rang, } (\ell \leq n). \qquad (36)$$

Les équations des conditions nécessaires, avec $\lambda_0 = 1$, s'écrivent

$$\dot{x} = (F-GR^{-1}S')x - GR^{-1}G'\lambda \qquad (37)$$

$$\dot{\lambda} = -(Q-SR^{-1}S')x - (F'-SR^{-1}G')\lambda \qquad (38)$$

et les conditions finales sur λ (conditions de transversalité)

$$\lambda(t_1) = Ax(t_1) + \Phi'\nu \qquad (39)$$

On voit alors que si X et Λ sont des matrices carrées satisfaisant (23), (24), (25), alors pour tout vecteur constant de ρ, de R^n

$$x(t) = X(t)\rho \quad , \tag{40}$$

$$\lambda(t) = \Lambda(t)\rho \quad , \tag{41}$$

sont solution de (36) à (39). En outre, si $X(t_0)$ est inversible, on peut coisir $\rho = X(t_0)^{-1}x_0$, satisfaisant ainsi aux conditions initiales.

Remarquons que $X(t_0)$ ne peut être inversible que si

$$\text{rang} \begin{bmatrix} X(t_1) \\ \Lambda(t_1) \end{bmatrix} = n \quad . \tag{42}$$

En effet, par la théorie des équations différentielles linéaires, ce rang est constant, quand t varie, et on veut qu'il soit égal à n à t_0.

La méthode proposée ici pour résoudre le problème à valeur limite en deux points (36) à (39) revient à ne prendre qu'une matrice de transition ($2n \times n$), et n'engendrant que les solutions satisfaisant aux conditions de transversalité. Si $\Phi = 0$ et $A = 0$, on prend précisément les n premières colonnes de la matrice de transition classique.

Notons la pseudo inverse de Φ par Φ^+

$$\Phi^+ = \Phi'(\Phi\Phi')^{-1} \text{ si } \Phi \neq 0, \text{ (et } \Phi^+ = 0 \text{ si } \Phi = 0)$$

un choix possible pour $X(t_1)$ et $\Lambda(t_1)$, dont on s'assurera qu'il satisfait aux conditions (21), (25) et (42), est

$$X(t_1) = I - \Phi^+\Phi \tag{43)a.$$

$$\Lambda(t_1) = A(I-\Phi^+\Phi) + \Phi^+\Phi \tag{44)a.$$

Alternativement, si le sous espace $\mathcal{U} = \{x | \Phi x = 0\} = \text{Ker}\Phi$ est engendré par la matrice de plein rang **M**, on a de manière équivalente

$$X(t_1) = MM^+ \tag{43)b.$$

$$(t_1) = AMM^+ + I - MM^+ \tag{44)b.$$

où $M^+ = (M'M)^{-1}M'$ est encore la pseudo-inverse.

REMARQUE Les expressions (40) et (41) de x et λ justifient l'emploi des notations X et Λ, qui satisfont bien ici encore à (19) et (20).

1.3.3 Equivalence des équations canoniques et de l'équation de Riccati

Dans cette brève section, nous considérons le cas à état final libre et démontrons le résultat ci-dessous [5], qui intervient dans le théorème 1.2.

THEOREME 1.3. L'équation de Riccati (30) a une solution sur $[t_0,t_1]$ si et seulement si il existe une solution aux équations (23), (24), (25) satisfaisant (21) et (22).

DEMONSTRATION Condition suffisante : Soit $X(t)$, $\Lambda(t)$ la solution de (23) à (25) satisfaisant (21) et (22). Introduisons

$$P(t) = \Lambda(t) \; X(t)^{-1} \tag{45}$$

Du fait que $X'\Lambda$ est symétrique pour tout t, on déduit que P(t) l'est aussi, car $P(t) = X'^{-1} (X'\Lambda) X^{-1})$. Par ailleurs, en dérivant (45), en utilisant (23) et (24) et $(X^{-1})^{\cdot} = - X^{-1}\dot{X} X^{-1}$, on vérifie directement que P(t) satisfait l'équation de Riccati, et (25) avec $\Phi = 0$ montre que $P(t_1) = A$.

Condition nécessaire. Soit P(t) solution de (30). Soit $\Phi_P(.,.)$ la matrice de transition associée à la matrice (dite du système bouclé)

$$F_P = F - GR^{-1}(G'P+S).$$

Posons

$$x(t) = \Phi_P(t,t_1)$$

$$\Lambda(t) = P(t)X(t)$$

On vérifie en dérivant ces deux expressions que X et Λ satisfont (23), (24). Comme $\Phi_P(t_1,t_1) = I$, $X(t_1) = I$, $\Lambda(t_1) = A$ satisfont (25) et (21). Enfin, toute matrice de transition est inversible, donc X(t) satisfait (22).

1.4 Commande en boucle fermée des systèmes non linéaires

Les ingénieurs préfèrent les lois de commande en boucle fermée (B.F.) ou stratégies $u^*(x,t)$ aux lois en boucle ouverte (B.O.) $u^0(x_0,t_0,t)$. En effet, les premières compensent mieux, en général, les perturbations toujours présentent dans la pratique. Cependant, la théorie d'Hamilton-Jacobi-Carathéodory est trop lourde pour être mise en oeuvre pour des systèmes non linéaires.

Dans [3], J.V. BREAKWELL et al. proposent de se contenter de minimiser la seconde variation de J autour d'une trajectoire optimale $u^0(t)$, $x_L(t)$ calculée à l'aide du Principe de Minimum de Pontryaguine. Grâce au caractère simple (L.Q.) de cette expression, cela est faisable et mène à la commande

$$\hat{u}(x,t) = u^0(t) - R^{-1}(G'P+S')(x-x^0) \qquad (46)$$

J.C. WILLEMS a fait remarquer le fait suivant (que la preuve ci-dessous permet immédiatement de généraliser au cas avec contraintes finales) :

THEOREME <u>L'expression</u> (46) <u>est le dévelopement au premier ordre de la stratégie optimale</u> $u^*(x,t)$ <u>au voisinage de la trajectoire nominale</u> $x^0(t)$.

DEMONSTRATION Par définition, $u\ (x\ (t),t) = u\ (t)$, de sorte qu'il suffit de vérifier que

$$\frac{\partial u^*}{\partial x} = -R^{-1}(G'P+S')$$

Avec les notations des § 1.2.1 et 1.2.2, on a, quand X est régulière et qu'on peut donc appliquer localement le théorème 1.2,

$$u^*(x,t) = \bar{u}(x,\ell(t,r(t,x)),t)$$

d'où

$$\frac{\partial u^*}{\partial x} = \frac{\partial u}{\partial x} + \frac{\partial u}{\partial \lambda}\frac{\partial \ell}{\partial \rho}\frac{\partial r}{\partial x}$$

Il ne reste plus qu'à utiliser les expressions de chacun des facteurs ci-dessus calculées au §1.2.2, et la relation (45) pour obtenir le résultat annoncé.

1.5. <u>Simplification des notations</u>

REMARQUE Dans tous les calculs relatifs au problème linéaire quadratique, on peut éliminer les termes en S par l'artifice suivant, consistant à faire un changement de variable de commande :

$$u = -R^{-1}S'x + \hat{u}$$

Le problème (27), (28) s'écrit alors

$$\dot{x} = \hat{F}x + G\hat{u} \quad , \quad x(t_0) = x_0 \quad ,$$

$$2J = x'(t_1)Ax(t_1) + \int_{t_0}^{t_1} (x'\hat{Q}x + \hat{u}'R\hat{u})\,dt \quad ,$$

avec

$$\hat{F} = F - GR^{-1}S' \quad , \quad \hat{Q} = Q - SR^{-1}S' \quad .$$

Toutes les équations subséquentes, notamment (30), (31), (37), (38), et donc aussi (23), (24) s'écrivent comme dans la forme originale, mais en y remplaçant F et Q par $\hat{F}$ et $\hat{Q}$, et en y faisant $S = 0$.

Nous nous limiterons donc, dans la suite, au cas $S = 0$ qui simplifie les écritures, sans perte de généralité. Nous reécrivons, à fin de références, les équations auxquelles cela nous mène :

$$\dot{P} + PF + F'P - PGR^{-1}G'P + Q = 0 \quad , \quad P(t_1) = A \tag{47}$$

$$u^* = -R^{-1}G'Px \tag{48}$$

et

$$\dot{X} = FX - GR^{-1}G' \quad , \quad X(t_1) = MM^+ \quad , \tag{49}$$

$$\dot{\Lambda} = -QX - F'\Lambda \quad , \quad \Lambda(t_1) = AMM^+ + I - MM^+ \tag{50}$$

$$\bar{u} = -R^{-1}G'\lambda \tag{51}$$

et (45) reste inchangée.

2. Normalité

2.1 Définitions et rappels

Nous définissons ce que nous appelons une <u>trajectoire anormale</u> du problème non linéaire (1), (3) de départ.

DEFINITION Une extrêmale du problème (1) (3) est dite anormale si elle satisfait au Principe du Minimum (eq. (4), (6)) avec $\lambda_0 = 0$.

REMARQUE Cette définition ne rattache qu'apparemment la notion d'extrêmale anormale au problème de commande optimale posé. En écrivant les conditions d'anormalité, on constate que c'est une propriété du seul système dynamique (1), (2).

Cette propriété est en fait une propriété de non commandabilité. C'est pourquoi nous faisons les rappels ci-dessous. Toutefois, nous nous limitons ici au cas où u est intérieur à U. (Rappelons qu'on a montré au §1.1.2 comment s'y ramener).

DEFINITION Le système (1) est localement complètement commandable de t_1, à t_0, le long de la trajectoire engendrée par (u(.) si la paire (F,G) correspondante est complètement commandable modulo $\mathcal{U}$ de t_1 à t_0.

C'est-à-dire qu'alors, le système (27) peut transférer tout état initial x_0 en $\mathcal{U}$ à l'instant t_1.

REMARQUE. Soit x_0^0, et $u^0(.)$ transférant l'état sur la cible :

$$\phi(x^0(t_1) = 0.$$

Soit $\mathcal{U}$ l'espace tangent à la cible en $x^0(t_1)$.

Ecrivons $x(t_1) = F(x_0,u(.))$, et $\Psi(x_0,u(.)) = \phi(F(x_0,u(.)))$.

La condition de complète commandabilité locale du système (1) est <u>équivalente</u> à la condition de fonction implicite pour l'existence locale d'une fonction $u(.) = \bar{u}(x_0)$ telle que

$$\Psi(x_0, \bar{u}(x_0)) = 0.$$

c'est-à-dire telle que $u(.) = \bar{u}(x_0)$ transfère x_0 sur la cible, pour tout x_0 dans un voisinage de x_0^0

Soit $W(t_0)$ l'ensemble des états transférables par le système (27) sur le sous espace donné $\mathcal{U}$ en t_1 donné aussi. On sait que

$$W(t_0) = \mathrm{Im}\,\phi(t_0,t_1)\,[\,MM^+ | \int_{t_0}^{t_1} \phi(t_1,s)\ G(s)\ G'(s)\ \phi'(t_1,S)\ ds\,] \quad (52)$$

où $\phi(.,.)$ désigne ici la matrice de transition associée à F et l'expression entre crochets la matrice formée en juxtaposant les deux blocs séparés par la barre.

On établit alors le résultat suivant :

THEOREME 2.1 Le fait, pour une trajectoire engendrée par une commande u (.) intérieure à U, d'être anormale est équivalent au fait que le système (1) ne soit pas localement complètement commandable.

DEMONSTRATION Le fait de n'être pas localement c-c est équivalent, naturellement, au fait que $W(t_0) \neq R^n$, et puisque $\phi(t_0,t_1)$ est régulière, au fait qu'il existe un vecteur ρ tel que

$$\rho' MM^+ = 0 \tag{53}$$

$$\rho' \int_{t_0}^{t_1} \phi(t_1,t)\, G(t)\, G'(t)\, \phi'(t_1,t)\, dt = 0$$

Mais en post multipliant la seconde relation par ρ, et en remarquant que $G'\phi'\rho$ est un vecteur continu, cette relation est équivalente à

$$\rho'\phi(t_1,t)\, G(t) = 0 \quad , \quad \forall t \in [t_0,t_1] \quad .$$

Ceci est équivalent, en posant

$$\lambda'(t) = \rho'\phi(t_1,t)$$

à

$$\dot{\lambda}' = -\lambda' F \quad , \quad \lambda'(t_1) = \rho' \tag{54}$$

$$\lambda'(t)\, G(t) = 0 \tag{55}$$

Mais (53) (54) est identique à (4) avec $\lambda_0 = 0$, et (55) à (6). Donc le théorème est établi. Nous utiliserons, dans la suite, (53) à (55) comme caractérisation d'un système non c.c. modulo $\mathcal{U}$.

2.2. Anormalité et champ d'extrêmales

L'objet de la première partie de ce chapitre est d'établir la deuxième propriété caractéristique des trajectoires anormales, qui est prise comme définition dans la littérature classique.

THEOREME 2.2 <u>Le fait, pour une trajectoire engendrée par u(.) intérieur à U, d'être anormale sur $[t_0,t_1]$ est équivalent au fait que la matrice X(t) engendrée par les équations canoniques correspondantes soit singulière sur tout cet intervalle.</u>

DEMONSTRATION i) Anormalité $\Rightarrow$ X(t) singulière. Notons que cette implication est naturelle, si on se souvient que X(t) engendre un champ de trajectoires qui aboutissent en $\mathcal{U}$. Si ce n'est pas tout R^n qui peut être transféré en $\mathcal{U}$, X(t) ne peut être régulière.

Spécifiquement, utilisons (49), (54) et (55) pour calculer

$$(\lambda'X)^{\cdot} = -\lambda'FX + \lambda'FX - \lambda'GR^{-1}G'\Lambda = 0$$

et (53) implique $\lambda'(t_1)X(t_1) = 0$, donc $\lambda'(t)X(t) = 0$ pour tout t. Q.E.D.
ii) X(t) singulière sur $[t_0,t_1]$ $\Rightarrow$ anormalité. Cette deuxième implication est beaucoup plus intéressante. Elle dit que si le champ des extrêmales a un plan tangent, ImX, différent de R^n, c'est qu'on se trouve à la frontière de l'ensemble des états qui peuvent être transférés sur la cible pour le problème initial.

Nous faisons d'abord la démonstration dans un cas particulier qui est intéressant en soi, permet une preuve très simple, et donne des résultats plus forts. Nous ferons ensuite la preuve dans le cas général.

a) cas $Q \geq 0$, $A \geq 0$.

Supposons alors $X(t_0)$ singulière (ce qui est beaucoup plus faible que l'hypothèse du théorème). Il existe donc un vecteur ρ tel que

$$X(t_0)\,\rho = 0. \tag{56}$$

Considérons le scalaire

$$v(t) = \rho'\Lambda'(t)\,X(t)\,\rho \quad .$$

Il satisfait aux relations (en utilisant (49) et (50) pour calculer $\dot{v}$)

$$v(t_0) = 0 \quad , \quad v(t_1) = \rho'MM^{+}AMM^{+}\rho \geq 0 \quad .$$

et

$$\dot{v}(t) = -\rho'X'QX\rho - \rho'\Lambda'GR^{-1}G'\Lambda\rho \leq 0 \quad .$$

De ces relations on déduit immédiatement

$$v(t) = 0 \quad \forall t \in [t_0,t_1] \quad ,$$

d'où en particulier $\dot{v} = 0$, et donc, R étant strictement positive

$$\rho'\Lambda'G = 0 \quad , \quad \forall t \in [t_0,t_1] \tag{57}$$

$$QX\rho = 0 \quad , \quad \forall t \in [t_0,t_1] \quad . \tag{58}$$

Posons $\lambda(t) = \Lambda(t)\rho$, en utilisant (50) et (58) il vient

$$\dot{\lambda} = - F'\lambda \quad . \tag{59}$$

Enfin, en utilisant (49) et (57) on a

$$\dot{(X\rho)} = F (X\rho) \quad ,$$

qui avec (56) permet de conclure

$$X(t)\, \rho = 0 \quad , \quad \forall t \in [t_0,t_1] \tag{60}$$

ce qui à l'instant t_1 donne

$$MM^+\rho = 0 \quad , \quad \text{d'où } \lambda(t_1) = \rho \quad . \tag{61}$$

Mais (57), (59) (61) sont identiques à (55) (54) (53), ce qui prouve le résultat dans ce cas a).

Comme notre seul hypothèse était (50), on en déduit le corollaire.

COROLLAIRE 0 Si la paire (F,G) est c.c. modulo $\mathcal{U}$ sur l'intervalle $[t_0,t_1]$, la matrice X(t) est régulière sur tout cet intervalle, sauf en t_1.

b) Le cas général

Pour établir l'implication ii) dans le cas féneral, nous prouvons d'abord deux lemmes, dont le second est très important.

LEMME 1 Le long d'une extrêmale, on a

$$\int_{t_0}^{t_1} (x'Qx + u'Ru)\, dt = \lambda'(t_0)\, x(t_0) - \lambda'(t_1)\, x(t_1) \quad .$$

PREUVE Il suffit de calculer la dérivée totale de $\lambda' x$ et d'utiliser (51).

LEMME 2 Il existe un intervalle (t_2,t_1) tel que si, pour $\tau_0 \in (t_2,t_1)$, et pour un vecteur $\rho \neq 0$, $X(\tau_0)\,\rho = 0$, alors $X(t)\,\rho = 0$ pour tout $t \in [\tau_0,t_1]$.

DEMONSTRATION Considérons le problème ayant même dynamique et même critère que le problème auxiliaire considéré, mais état final libre. On lui associe des équations canoniques, dont nous noterons $\bar{X}$, $\bar{\Lambda}$ la solution. On a $\bar{X}(t_1) = I$, et donc, par continuité, il existe un intervalle (t_2,t_1) sur lequel $\det \bar{X} \neq 0$. Sur cet intervalle, on peut appliquer la théorie des conditions suffisantes à ce nouveau problème. Soit en particulier $\tau_0 \in (t_2,t_1)$ et $x(\tau_0) = 0$. Alors on peut affirmer que la trajectoire $y(t) = 0$, $x(t) = 0$ donne un coût $J = 0$, et que toute autre trajectoire donne un coût strictement positif.

Revenons maintenant au problème à état final contraint, et supposons que pour $\tau_0 \in (t_2,t_1)$, $X(\tau_0)\,\rho = 0$. Considérons l'extrêmale $x(t) = X(t)\,\rho$, et appliquons lui le lemme 1. Il vient

$$J = x'(t_1)\ Ax(t_1) + \int_{\tau_0}^{t_1} (x'Qx + u'Ru)\,dt = \lambda'(\tau_0)\ x(\tau_0) = 0.$$

Mais d'après l'argument précédent, de toutes les trajectoires issues de 0 à τ_0, (et quelque soit, par ailleurs, leur état final), seule la trajectoire $x(t) = 0$ peut annuler J, d'où le résultat.

En conséquence du lemme 2, ont peut affirmer que si $X(t)$ est singulière sur $[t_0,t_1]$, il existe un instant t_2 et un vecteur fixe ρ tel que

$$X(t)\,\rho = 0 \quad , \quad \forall t \in [t_2,t_1] \quad . \tag{62}$$

Dérivons en utilisant (49), et (62) à nouveau, il vient

$$G'(t)\ \Lambda(t)\ \rho = 0 \quad \forall t \in [t_2,t_1] \quad .$$

En posant $\lambda(t) = \Lambda(t)\,\rho$, et en utilisant (62) dans le calcul de $\dot{\lambda}$ et directement en t_1, on retrouve (57), (59) et (61), et on conclut que (F,G) n'est pas c.c. modulo $\mathcal{U}$ sur $[t_2,t_1]$.

Supposons maintenant que (F,G) soit c.c. modulo sur $[t_0,t_1]$, et appelons t_2, encore, le plus petit instant tel que $W(t_2) \neq R^n$, qui existe, du fait de la continuité des déterminants de la matrice (52). Posons $A_2 = \Lambda(t_2)\, X^+(t_2)$, et considérons le problème avec même dynamique, instant final t_2, même coût intégral, coût final $x'A_2x$, et état final contraint d'appartenir à $\mathcal{U}_2 = \operatorname{Im}X(t_2)$.

Alors on peut initialiser les équations canoniques de ce problème avec $X(t_2)$ et $\Lambda(t_2)$ puisque, d'une part, on a bien $\operatorname{Im}X(t_2) = \mathcal{U}_2$, et d'autre part on a, en écrivant X_2, Λ_2 pour $X(t_2)$, $\Lambda(t_2)$,

$$\Lambda_2 = \Lambda_2 X_2^+ X_2 + \Lambda_2(I - X_2^+ X_2) \quad ,$$

soit

$$\Lambda_2 = A_2 X_2 + \Lambda_2(I - X_2^+ X_2) \quad ,$$

et le dernier terme ci-dessus engendre un sous espace orthogonal à , comme on le vérifie en prémultipliant par x_2' et en utilisant (21) :

$$X_2'\Lambda_2(I - X_2^+ X_2) = \Lambda_2' X_2 (I - X_2^+ X_2) = 0 \quad .$$

Donc ce nouveau problème admet les mêmes matrices $X(t)$,$\Lambda(t)$ pour solution de ses équations canoniques. Comme $X(t)$ est par hypothèse singulière sur $[t_0,t_1]$, on conclut par application du lemme ci-dessus à l'existence d'un instant $t_3 < t_2$ tel que la paire (F,G) ne soit pas c.c. modulo $\mathcal{U}$ sur $[t_3,t_2]$. On va montrer ci-dessous (indépendamment de cet argument de continuation), que $\mathcal{U}_2 = W(t_2)$, donc la paire (F,G) n'est pas c.c. modulo $\mathcal{U}$ sur $[t_3,t_1]$, ce qui contredit l'hypothèse que $W(t) = \mathbb{R}^n$ pour tout $t < t_2$. Le théorème est démontré.

On a, au passage, établi le résultat suivant, qui est très important :

COROLLAIRE 1 Il existe un intervalle non vide (t_2,t_1) sur lequel le noyau Ker X(t) est strictement croissant, donc constant par morceaux.

Le caractère croissant de Ker X(t) n'est qu'une autre façon d'exprimer le lemme 2. Et comme ce noyau est un sous espace vectoriel, la seule façon dont il puisse croître est en gagnant une dimension, ce qui ne peut arriver qu'un nombre fini (<n), de fois. Remarquons qu'on a prouvé que l'intervalle pour lequel cette propriété est vraie contient

au moins celui sur lequel $\overline{X}(t)$ est inversible. Mais il peut être strictement plus grand.

2.3 Etude du cas anormal

L'objet de ce paragraphe est de ramener l'étude de cas anormal à celle du cas normal, pour lui appliquer ensuite les résultats standards. Auparavant, nous montrons une généralisation des formules (29) (45) et (48).

THEOREME 2.3 Dans le cas anormal, on a encore (29) et (48), si on pose

$$P = \Lambda X^{+} \tag{45b}$$

DEMONSTRATION Considérons une trajectoire engendrée par ρ donné, et plaçons nous dans un intervalle où Ker $X(t)$ est constant. On a $x(t) = X(t)\,\rho$, $\lambda(t) = \Lambda(t)\,\rho$, $\overline{u}(t) = -R^{-1}G'\lambda(t)$. On peut écrire

$$\rho = X^{+}X\rho + (I-X^{+}X)\,\rho = X^{+}x + \sigma$$

σ est la projection orthogonale de ρ sur Ker X. Comme Ker X est constant, σ est un vecteur constant de Ker X. On a donc

$$0 = \dot{(X\sigma)} = -GR^{-1}G'\Lambda\sigma$$

soit $G'\Lambda\sigma = 0$. En reportant dans $\overline{u}$, il vient, avec la définition (45b)

$$\overline{u} = -R^{-1}G'X^{+}x - R^{-1}G'\Lambda\sigma = -R^{-1}G'Px \quad .$$

De plus, d'après le lemme 1 on sait que

$$V(x,t) = x'(t)\,\lambda(t) = x'\Lambda X^{+}x + \rho'X'\Lambda\sigma = x'Px + \rho'\Lambda'X\sigma = x'Px$$

en utilisant la symétrie de $X'\Lambda$. Enfin, à un instant t_2 où Ker X change, on peut calculer le nouveau σ. On sait que dans un voisinage gauche il reste constant, et les relations ci-dessus se prolongent par continuité à l'instant t_2. On vérifiera que cela n'introduit pas de discontinuité dans Λx^{+}.

THEOREME 2.4. La matrice $X(t)$ engendre un champ de trajectoires régulier dans le sous espace $W(t)$ des états commandables modulo $\mathcal{U}$ à t_1. (Muni de la topologie naturelle induite par R^n).

Nous nous plaçons dans un intervalle (t ,t) où W(t) est de dimension constante (qui existe, puisque la dimension de W ne peut aller que décroissante). Il existe une base, fixe, où la matrice

$$\Gamma(t_2) = [MM^+ | \int_{t_2}^{t_1} \phi(t_1,s)\ G(s)\ G'(s)\ \phi'(t_1,s)\ ds]$$

et la matrice M s'écrivent

$$\Gamma = \begin{pmatrix} \Gamma_1 \\ 0 \end{pmatrix} , \quad M = \begin{pmatrix} M_1 \\ 0 \end{pmatrix} , \quad \Gamma_1 \text{ de plein rang ;}$$

On vérifie facilement qu'il en va de même pour $\Gamma(t)$, $\forall t \in [t_2,t_1]$ dans la même base. Et W(t) est l'image de $\phi(t,t_1)\ \Gamma(t)$.

Etudions les matrices

$$Z(t) = \phi(t_1,t)\ X(t) \quad \text{ou} \quad X(t) = \phi(t,t_1)\ Z(t)$$

$$Y(t) = \phi'(t,t_1)\ \Lambda(t) \quad \text{ou} \quad \Lambda(t) = \phi'(t\ ,t)\ Y(t)$$

En posant

$$\tilde{G} = \phi(t_1,t)\ G(t) = \begin{pmatrix} G_1(t) \\ 0 \end{pmatrix} , \quad \tilde{Q} = \phi'(t,t_1)\ Q\phi(t,t_1)$$

on obtient immédiatement

$$\dot{Z} = -\tilde{G}R^{-1}\tilde{G}'Y$$

$$\dot{Y} = -\tilde{Q}Z$$

Ecrivons ces équations bloc par bloc pour la partition induite par celle de Γ :

$$\dot{Z}_{11} = -G_1R^{-1}G_1'Y_{11} \qquad Z_{11}(t_2) = M_1M_1^+$$

$$\dot{Z}_{21} = 0 \qquad Z_{21}(t_1) = 0$$

$$\dot{Y}_{11} = \tilde{Q}_{11}Z_{11} - \tilde{Q}_{12}Z_{21} , \qquad Y_{11}(t_1) = A_{11}M_1M_1^+ + I - M_1M_1^+ .$$

D'où il ressort Tu Z_{11} et Y_{11} satisfont les équations canoniques du système

$$\dot{z}_1 = G_1 u \quad , \quad J = z_1'(t_1) A_{11} z_1(t_1) + \int_{t_0}^{t_1} (z_1' \tilde{Q}_{11} z_1 + u'Ru)\, dt \quad .$$

Mais Γ, étant de plein rang, le système est c.c. modulo $\mathcal{U}$ à t_1. Donc d'après le théorème 2.2, Z_{11} est de plein rang sur tout l'intervalle. Donc $X(t)$ engendre tout $W(t)$.

Enfin, aux instants où W change de dimension, on applique le même raisonnement au problème à instant final t_2, $\mathcal{U}_2 = W(t_2)$, et à coût final $x'P(t_2)\,x$ (où P est, bien sur, défini par (45b)). Ceci permet d'étendre le résultat à t_0, t_1 .

En revenant au problème non linéaire d'origine, dont le problème linéaire quadratique est le problème auxiliaire, cela veut dire que les extrêmales voisines de celle étudiée forment une variété dont $X(t)$ engendre l'espace tangent, et que le champ, restreint à cette variété est régulier. On peut donc lui appliquer la théorie du cas normal et obtenir ainsi l'optimalité des trajectoires anormales.

En conséquence, à partir de maintenant nous restreignons notre étude au cas normal, ou c.c. modulo $\mathcal{U}$ sur t, t_1 , $\forall t \in [t_0, t_1]$.

3. Points focaux et optimalité

L'objet de ce chapitre est d'étudier ce qui se passe aux instants où $X(t)$ cesse d'être inversible, que ce soit t , à cause d'une contrainte finale, ou le plus grand instant $t_1^* < t$ pour lequel cela arrive, puisque jusqu'ici notre étude est limitée au cas $t_0 \in (t_1^*, t_1)$.

3.1 Intégrale complète

Afin de replacer la démonstration du théorème 3.1 dans son contexte, et pour l'intérêt de pousser plus loin le parallèle entre la théorie de l'équation de Riccati et des équations linéaires associées, d'une part et celle des équations canoniques d'autre part, nous développons ce paragraphe un peu plus que strictement nécessaires pour notre objet.

Nous avons vu que sur tout intervalle où elle est définie, la fonction $V(x,t) = \frac{1}{2} x'P(t)x$ est une solution de l'équation d'Hamilton Jacobi, ceci dès que P satisfait à l'équation de Riccati, et quelle que soit la condition finale. C'est cette condition que nous allons faire varier.

Une intégrale complète est une fonction $V(x,t;\sigma)$, dépendant des n paramètres σ_i, solution de l'équation d'Hamilton Jacobi pour toute valeur fixe de σ, et telle que le gradient $V_x(x,t_1;\sigma)$ engendre, pour tout x, tout R^n quand σ varie.

(On modifiera légèrement cette définition par la suite).

Considérons deux solutions des équations canoniques, l'une indicée par ρ l'autre par σ, et définies par leurs conditions finales

$$X_\rho(t_1) = I \quad , \quad X_\sigma(t_1) = 0 \quad ,$$

$$\Lambda\rho(t_1) = A \quad , \quad \Lambda_\sigma(t_1) = I \quad . \tag{63}$$

La solution ρ correspond au problème à état final libre, et la solution σ au champ distingué de t_1. Ces 2n solutions vectorielles des équations canoniques engendrent toutes les solutions de ces équations.

Prenons donc deux vecteurs ρ et σ fixés, et considérons la trajectoire

$$x(t) = X_\rho(t)\rho + X_\sigma(t)\sigma$$

$$\lambda(t) = \Lambda_\rho(t)\rho + \Lambda_\sigma(t)\sigma$$

A σ fixé, en faisant varier ρ on a encore un quasichamp, car on a encore

$$\left[\frac{\partial x(t_1)}{\partial\rho}\right]' \left[\frac{\partial\lambda(t_1)}{\partial\rho}\right] = A = \left[\frac{\partial\lambda(t_1)}{\partial\rho}\right]' \left[\frac{\partial x(t_1)}{\partial\rho}\right] .$$

On peut alors effectuer la substitution :

$$\rho = X^{-1}(x - X_\sigma\sigma) \quad ,$$

$$\lambda = \Lambda_\rho X_\rho^{-1}x + (\Lambda_\sigma - \Lambda_\rho X_\rho)\,\sigma = Px + \Gamma\sigma \quad .$$

Nous avons posé

$$\Lambda_\rho X_\rho^{-1} = P \quad , \quad \Lambda_\sigma - PX_\sigma = \Gamma \ . \tag{64}$$

Introduisons encore la matrice

$$\Pi = - \Gamma' X_\sigma = - X_\sigma' \Gamma \quad . \tag{65}$$

On vérifie directement que ces matrices satisfont, pour P l'équation de Riccati naturellement, et pour Γ et Π :

$$\dot{\Gamma} + (F' - PGR^{-1}G')\,\Gamma = 0 \quad , \quad \Gamma(t_1) = I \quad , \tag{66}$$

$$\dot{\Pi} - \Gamma' G R^{-1} G' \Gamma = 0 \quad , \quad \Pi(t_1) = 0 \quad , \tag{67}$$

et aussi que la fonction ci-dessous est une intégrale complète :

$$2V(x,t;\sigma) = x'Px + x'\Gamma\sigma + \sigma'\Gamma'x + \sigma'\Pi\sigma$$

satisfaisant

$$2V(x,t\ ;\sigma) = x'Ax + 2\sigma'x \quad . \tag{68}$$

Ici, la commande optimale associée, qui minimise l'hamiltonien, est

$$u^* = - R^{-1}G'(Px + \Gamma\sigma) \quad . \tag{69}$$

On vérifie aussi que les équations (66) et (67) permettent d'identifier $g = \Gamma\sigma$ et $h = \sigma'\Pi\sigma$ avec les solutions de (34) (35) introduites dans la première partie, et que (68) et (69) sont bien identiques à (33).

On peut d'ailleurs vérifier que

$$\Gamma = X_\rho'^{-1} \quad , \quad \Pi = - X_\rho^{-1} X_\sigma \quad .$$

Soit en effet

$$\Psi = \begin{pmatrix} X_\rho X_\sigma \\ \Lambda_\rho \Lambda_\sigma \end{pmatrix}$$

la matrice de transition complète associée aux équations canoniques (49), (50). Le résultat suivant est célèbre :

PROPOSITION. La matrice Ψ est symplectique, c'est-à-dire

$$\Psi' J \Psi = J ,$$

où

$$J = \begin{bmatrix} 0 & I \\ -I & 0 \end{bmatrix}$$

DEMONSTRATION. La matrice Ψ satisfait l'équation différentielle

$$\dot{\Psi} = \mathcal{H} \Psi$$

où $\mathcal{H}$ est la matrice du système canonique, qui satisfait, comme toute matrice des dérivées partielles d'un système hamiltonien

$$\mathcal{H}' J + J \mathcal{H} = 0 .$$

Il suffit alors de calculer

$$(\Psi' J \Psi)^{\cdot} = \Psi' (\mathcal{H}' J + J \mathcal{H}) \Psi = 0$$

et de noter que $\Psi(t_0,t_0) = I$ pour obtenir le résultat annoncé.

Le caractère symplectique de Ψ s'écrit ici

$$\Lambda'_\alpha X_\alpha - X'_\alpha \Lambda_\alpha = 0 , \quad \alpha = \rho,\sigma$$

$$\Lambda'_\sigma X_\rho - X'_\sigma \Lambda_\rho = I .$$

La première relation est celle du lemme 1.1., la seconde donne Γ en fonction de $X\rho$.

3.2. La construction de Tonelli

3.2.1. Le problème non linéaire

L'objet de ce parahraphe est d'étendre le théorème de Carathéodory au cas où, la cible étant non triviale, la fonction V construite par la méthode des caractéristiques n'est pas C^1 en t_1.

En effet, en chaque point de la cible, i.e. chaque x tel que

$$\phi(x) = 0 \quad ,$$

on a une infinité de $\lambda(t_1)$ possibles, qui sont données par

$$\lambda(t_1) = \frac{\partial K}{\partial x}(x) + \sum_{i=1}^{p} \nu_i \frac{\partial \phi_i}{\partial x}(x) \quad ,$$

pour tout ν_i, et qui engendrent chacun une trajectoire distincte arrivant en x. Comme on a vu que λ était le gradient V_x de x, il est clair que la fonction V ainsi construite n'est pas continûment dérivable en x à l'instant t_1, puisqu'il n'y a pas unicité de la limite $\lambda(t_1)$ du gradient $\lambda(t)$ pour diverses trajectoires arrivant au même point x.

On va montrer que le champ des trajectoires ainsi construites est néanmoins optimal dans tout voisinage de la cible où il est régulier.

Considérons un état final $\bar{x}_1$ donné, et un adjoint $\bar{\lambda}_1$ donné par un $\bar{\nu}$ fixé, et la trajectoire $\bar{x}(t)$, $\bar{\lambda}(t)$, solution des équations différentielles (1) et (4), associée.

A cette trajectoire, nous associons un nouveau problème d'optimisation, ayant même dynamique et même critère que celui d'origine, mais à instant final libre, et avec pour cible

$$\phi(x,t) = 0$$

où

$$\bar{\phi}(x,t) = \Sigma \bar{\nu}_i \phi_i(x) - (t-t_1)\bar{H}_1 \quad ,$$

$\bar{H}_1$ étant la valeur de l'hamiltonien optimal en $\bar{x}_1$, $\bar{\lambda}_1$.

On remarque que la cible d'origine est contenue dans la nouvelle cible.

Ce nouveau problème a une cible de dimension n dans l'espace des phases (x,t) de dimension n + 1. Pour cette raison, les conditions de transversalité (auxquelles il faut ajouter $- \bar{H} = 0$ à l'instant final t_f) définissent un $\lambda(t_f)$ et un seul en chaque phase (x,f,t_f) de la cible.

On peut donc, localement, construire un champ de caractéristiques de l'équation de Hamilton Jacobi de Carathéodory associée, régulier jusque sur la cible, auquel le théorème de Carathéodory s'applique.

Cette cible a été choisie de telle façon que $\bar{\lambda}_1$ soit l'adjoint solution des conditions de transversalité étendues, en $\bar{x}_1$. Donc la trajectoire $\bar{x}(t)$ est optimale pour le nouveau problème. Mais comme toute trajectoire de comparaison du problème d'origine est a fortiori une trajectoire de comparaison du nouveau problème (car les cibles sont emboîtées), et que le coût est le même dans les deux, la trajectoire $\bar{x}(t)$ est donc aussi optimale pour le problème d'origine.

Cet argument est correct dans un voisinage de la nouvelle cible où le champ associé est régulier. La théorie de l'intégrale complète permet de montrer l'existence d'un voisinage V de la cible d'origine contenu dans tous ces voisinages. Ainsi, le champ construit dans le problème d'origine est optimal dans V , mais également à l'extérieur de V aussi longtemps qu'il est régulier, comme on le voit en appliquant un argument de programmation dynamique.

Remarquons que cet argument est nécessaire pour établir complètement le théorème 1.2 dans le cas d'une cible non triviale. Nous le particularisons maintenant au cas linéaire quadratique, qui va nous montrer comment la connaissance d'une intégrale complète permet de donner la solution de l'équation d'Hamilton Jacobi Carathéodory associée au nouveau problème de chaque trajectoire.

3.2.2. Le problème linéaire quadratique

THEOREME 3.1. <u>Les équations canoniques (49) (50) engendrent des trajectoires optimales pour le problème linéaire quadratique correspondant sur tout intervalle sur lequel</u> $X(t)$ <u>est inversible sauf en</u> t_1.

DEMONSTRATION On modifie l'intégrale complète ci-dessus en prenant pour conditions finales sur X_σ, Λ_σ celles de (49) (50). Cette nouvelle intégrale complète est associée au coût final

$$K(x) = \frac{1}{2} x'Ax + \sigma'(I - MM^+) x \tag{70}$$

Donc, pour tout $x \in \mathcal{U}$, ce coût final coïncide avec celui du problème homogène.

A cause de la condition finale $X\rho(t_1) = I$, il existe un voisinage (t_2,t_1) où le champ de trajectoires associé à cette intégrale complète est régulier, quelque soit σ. De plus, dans cet intervalle, $X_\sigma(t)$ est inversible, à cause du théorème 2.2 et du lemme 2. Considérons le problème à état final contraint, et commençant à l'instant t_2

Soit $x(t_2) = x_2$ l'état initial, et $\overline{\sigma}$ tel que

$$X_\sigma(t_2)\,\overline{\sigma} = x_2 \quad .$$

La trajectoire

$$x(t) = X_\sigma(t)\,\overline{\sigma} \quad , \tag{71}$$

qui est celle dont nous devons montrer l'optimalité, est une extrémale associée à la nouvelle intégrale complète $V(x,t;\overline{\sigma})$, avec $\rho = 0$. Donc elle minimise le critère quadratique avec coût final (70). Donc toute autre trajectoire donne une valeur plus grande à ce critère, en particulier toute autre trajectoire arrivant en $\mathcal{U}$ à l'instant t_1. Mais pour celles-là le nouveau critère coïncide avec l'ancien. Donc la trajectoire (71) est bien optimale sur (t_2,t_1) pour le problème linéaire quadratique homogène à état final contraint.

Maintenant, nous pouvons considérer le problème sur $[t_0,t_2]$, avec état final libre en t_2 et coût $V(x,t_2)$. Sur cet intervalle $X(t)$ (qui coïncide avec $X_\sigma(t)$) est inversible et on dispose de la théorie élémentaire pour prouver l'optimalité de la trajectoire considérée.

3.3 Points conjugués

DEFINITION On appelle point conjugué (arrière) de $(t_1, \mathcal{U})$ le plus grand instant $t_1^* < t_1$ tel que $X(t_1^*)$ soit singulier.

Cette définition n'a de sens que dans le cas normal, auquel nous supposons toujours nous être ramenés. Autrement il faudrait se référer au premier instant où le rang de $X(t)$ diminue.

Le corollaire 1 du lemme 2 et le corollaire 0 du théorème 2.2 nous donnent des résultats d'existence de l'intervalle (t_1^*,t_1) :

THEOREME 3.2 Les points conjugués sont isolés. Si A et Q sont positives semi-définies, il n'y a pas de point conjugué arrière de t_1, quelque soit $\mathcal{U}$.

REMARQUE importante. Dans toute cette théorie, le sens du temps est peu important. On aurait pu, comme Carathéodory, considérer le problème à état initial dans $\mathcal{U}_1$, et discuter suivant l'état final $x(t_0) = x_0$ imposé, pour $t_o^1 > t^1$. Ainsi tous les résultats se transposent en temps positif : optimalité du champ jusqu'à $\mathcal{U}$, point conjugué avant isolé. Remarquons en particulier le fait suivant :

PROPOSITION S'il existe un point conjugué arrière t_1^* de $(t_1, \mathcal{U})$, alors t_1 est le point conjugué avant de $(t_1^*, \mathrm{Im} X(t_1^*))$.

La preuve est immédiate et consiste à vérifier que $X(t_1^*)$, $\Lambda(t_1^*)$ constituent la bonne initialisation des équations canoniques du problème avant. Ceci se fait avec le même argument que dans la preuve du théorème 2.2.

En adaptant de façon évidente la preuve du lemme 2, on a le résultat suivant :

THEOREME 3.3. Si on considère deux problèmes linéaires quadratiques ne différant que par leurs cibles finales $\mathcal{U}_1$ et $\mathcal{U}_2$, et si ces cibles sont emboîtées : $\mathcal{U}_1 \supset \mathcal{U}_2$, et si on appelle t_1^* et t_2^* les points conjugués arrières de t_1 correspondants, alors $t_1^* \geq t_2^*$.

DEMONSTRATION Comme lemme 2, en comparant le problème 2 au problème 1, au lieu de $\mathcal{U} = R^n$. Il nous reste un résultat, fondamental, à établir (cf. [2]).

THEOREME 3.4 Il n'existe pas de solution au problème linéaire quadratique sur un intervalle contenant un point conjugué dans son intérieur.

DEMONSTRATION Les équations canoniques constituent des conditions nécessaires d'optimalité. Remarquons que pour $t_0 < \overset{*}{t}_1$, $X(t_0)$ est de nouveau inversible (sauf si t_0 est le deuxième point conjugué). Soit ρ_0 tel que $x_0 = X(t_0)\,\rho_0$. Seule la trajectoire $x(t) = X(t)\,\rho_0$ peut être optimale.

A l'instant t_1, on a, comme dans la démonstration du lemme 2,

$$\Lambda(t_1) = A_1 X(t_1) + N_1$$

où $A_1 = \Lambda(t_1)\, X(t_1)^+$, et N_1 engendre un sous espace vectoriel normal à $\mathrm{Im}X(t_1)$. Soit σ dans le noyau de $X(t_1)$. On a :

$$X(t_1)\sigma = 0$$

$$\Lambda(t_1)\sigma = N_1\sigma$$

Comparons les trajectoires engendrées par ρ_0 et par $\rho_0 + \sigma$. Elles ont même état $x(t_1)$. En appliquant le lemme 1, on trouve qu'elles donnent la même valeur au coût entre $\overset{*}{t}_1$ et t_1. En effet, pour $\rho_0 + \sigma$, on a

$$J[\overset{*}{t}_1, t_1] = \lambda'(\overset{*}{t}_1)\; x(\overset{*}{t}_1) = \rho_0'\Lambda'(\overset{*}{t}_1)\; x(\overset{*}{t}_1) + \sigma' N_1' x(\overset{*}{t}_1)$$

et le dernier terme est nul à cause de la propriété de N_1.

Ainsi, depuis t_0, on peut considérer la trajectoire définie par

$$x(t) = X(t)\,\rho_0 \quad , \quad t_0 \leq t \leq \overset{*}{t}_1$$

$$x(t) = X(t)(\rho_0 + \sigma) \quad , \quad t_1 \leq t \leq \overset{*}{t}_1 \; .$$

Cette trajectoire donne la même valeur à J que celle de référence, mais elle ne satisfait pas les conditions nécessaires d'optimalité. D'où le résultat.

En utilisant le théorème 3.3, on pourrait choisir un instant $t_2 < \overset{*}{t}_1$ tel qu'en construisant un champ avant depuis $\mathcal{U}_2 = \{a x(t_2)\}$, $a \in R$, il existe un instant $t_3 < \overset{*}{t}_1$ où ce champ est régulier, et on exhiberait ainsi une trajectoire utilisant ce champ entre t_2 et t_3, et les points du champ $X(t)$ qu'elle connecte, qui donnerait à J une valeur inférieure à $\lambda'(t_0)\; x(t_0)$.

Si on revient au problème non linéaire initial, cela veut encore dire qu'une extrémale contenant un point conjugué n'est pas optimale. Cela ne dit rien, bien sûr, quant à l'existence d'une autre solution au même problème, puisque la seconde variation est une étude purement locale.

BIBLIOGRAPHIE

[1] P. BERNHARD, Commande Optimale, Décentralisation et Jeux Dynamiques. Dunod, 1976.

[2] J.V. BREAKWELL & Y.C. HO "On the Conjugate Point Condition for the Control Problem", Int. J. of Engineering Sc., 2, pp. 565-579, 1965.

[3] J.V. BREAKWELL, J.L. SPEYER & A.E. BRYSON, "Optimization and Control of Nonlinear Systems using the Second Variation", SIAM J., 1, 1963.

[4] C. CARATHEODORY, Calculus of Variations and Partial Differential Equations of the First Order, (traduction Holden Day, 1962) Original en allemand, 1923.

[5] R.E. KALMAN, "Contribution to the Theory of Optimal Control", Bull. Soc. Math. Mexicana, 1960.

[6] N.N. PONTRYAGUINE, et al., Mathematical Theory of Optimal Processes, Interscience, New York, 1962.

-:-:-:-:-:-

METHODES ASYMPTOTIQUES DANS L'ETUDE DE SYSTEMES HAMILTONIENS NON AUTONOMES.

J. BLOT

CEREMADE.

METHODES ASYMPTOTIQUES DANS L'ETUDE DE SYSTEMES HAMILTONIENS NON AUTONOMES

J. BLOT
CEREMADE.

Résumé : On étudie une famille de systèmes hamiltoniens perturbés par un petit paramètre ε . On se donne une solution T-périodique du système non perturbé, et on cherche ce qu'elle devient pour des valeurs $\varepsilon > 0$ du paramètre. On donne deux exemples et un cas de bifurcation.

Abstract : We study a family of Hamiltonian systems depending on a small parameter ε . We are given a T-periodic solution of the unperturbed system, and we investigate what happens to it for positive values of ε . We give two examples and a case of bifurcation.

§ I : SUR LA TRANSFORMEE DE LEGENDRE

Soient N,M deux entiers strictement positifs ; on considère les applications $H : R \times R^{2N} \times R^M \to R$ et $f : R \to R^{2N}$ telles que :

[hyp. 1]
$$\begin{cases} . \ H \text{ est } C^{r+1} \quad (r \geq 2) \\ . \ \text{pour tout } (t,u,\varepsilon) \in R \times R^{2N} \times R^M, \quad H(t+T,u,\varepsilon) = H(t,u,\varepsilon) \\ . \ f \text{ est T-périodique et } C^1 \quad (T > 0) \end{cases}$$

en notant $u = (q,p) \in R^N \times R^N = R^{2N}$, on considère les équations hamiltoniennes perturbées forcées, ε jouant le rôle des paramètres de perturbation, $f = (f_1,f_2)$ jouant le rôle de la force extérieure :

$$\begin{cases} \dot{q} = \dfrac{\partial H}{\partial p}(t,q,p,\varepsilon) + f_1(t) \\ \dot{p} = -\dfrac{\partial H}{\partial q}(t,q,p,\varepsilon) + f_2(t) \end{cases}$$

en introduisant la matrice symplectique $\sigma = \begin{pmatrix} 0 & id_{R^N} \\ -id_{R^N} & 0 \end{pmatrix}$ qui satisfait $t_\sigma = \sigma^{-1} = -\sigma$, ces équations s'écrivent

$$(H_\varepsilon) \qquad \dot{u} = \sigma H'_u(t,u,\varepsilon) + f(t)$$

pour $\varepsilon = (\varepsilon^1,\ldots,\varepsilon^M) \in (R + \star)^M$, on notera $P(\varepsilon)$ le polydisque $\{(r^1,\ldots,r^M) \in R^M \ / \ |r^k| < \varepsilon^k \ ; \ 1 \leq k \leq M\}$; $D(u,R)$ désigne la boule ouverte de centre u et de rayon R dans R^{2N}.

<u>THEOREME 1</u> : Sous [hyp. 1] et l'hypothèse

[hyp. 2]
$$\begin{cases} \text{il existe une solution T-périodique, } t \to u_0(t), \text{ de } (H_0) \text{ et} \\ \text{pour tout } t \in R : H''_{uu}(t,u_0(t),0) \text{ est définie positive} \end{cases}$$

il existe $a > 0$, $\alpha > 0$, $\tilde{\varepsilon} \in]0,\infty[^M$ tels que en notant $W(\alpha) = \{(t,u) \in R \times R^{2N} \; / \; |u-u_o(t)| < \alpha\}$ on ait : pour tout $(t,u,\varepsilon) \in W(\alpha) \times P(\tilde{\varepsilon})$, pour tout $\delta u \in R^{2N}$, $H''_{uu}(t,u,\varepsilon)(\delta u,\delta u) \geqslant a|\delta u|^2$.

PREUVE : Considérons l'application $(t,\delta u) \to H''_{uu}(t,u_o(t),0)(\delta u,\delta u)$; elle est continue et strictement positive en chaque point du compact $[0,T] \times S^{2N-1}$ (S^{2N-1} = la sphère unité de l'espace R^{2N}) d'après [hyp. 2]. Un théorème de Weierstrass nous permet d'affirmer que cette fonction atteint son minimum sur ce compact, donc il existe $\tilde{t}$, $\tilde{\delta u}$ tels que $H''_{uu}(t,u_o(t),0)(\delta u,\delta u) \geqslant H''_{uu}(\tilde{t},u_o(\tilde{t}),0)(\tilde{\delta u},\tilde{\delta u}) = m > 0$. En outre, l'application H''_{uu} est continue de $R \times R^{2N} \times R^M$ dans $L_2(R^N,R^N;R)$ donc continue en chaque point $(t,u_o(t),0)$, ainsi :

pour tout $t \in [0,T]$ il existe Ω_t , voisinage ouvert de $(t,u_o(t))$ dans $R \times R^{2N}$, il existe $\varepsilon_t^k > 0$ $(1 \leqslant k \leqslant M)$ tels que :

$$(t',u) \in \Omega_t \; , \; \varepsilon \in P(\varepsilon_t) \Rightarrow \|H''_{uu}(t',u,\varepsilon)-H''_{uu}(t,u_o(t),0)\| \leqslant \frac{m}{4}$$

ici $\varepsilon_t = (\varepsilon_t^1,\ldots,\varepsilon_t^M)$ et la norme est la norme des formes bilinéaires continues.

Remarquons que nous pouvons prendre Ω_t aussi petit que nous le désirons, ainsi par définition de la topologie produit nous pouvons prendre Ω_t de la forme $]t-s_t \, , \, t+s_t[\times D(u_o(t),R_t)$ (D désignant la boule ouverte), et puisque u_o est continue, il existe $r_t > 0$, $(r_t \leqslant s_t)$ tel que $u_o(]t-r_t \, ; \, t+r_t\,) \subset D(u_o(t),R_t)$. Nous choisirons $\Omega_t =]t-r_t \, , \, t+r_t[\times D(u_o(t),R_t)$.

Notons $\gamma = \{(t,u_o(t)) \; / \; t \in [0,T]\}$ c'est un compact et les Ω_t, lorsque t décrit $[0,T]$, constituent un recouvrement ouvert de γ , donc en vertu du théorème de Heine-Borel-Lebesgue, on peut en extraire un sous-recouvrement fini : Ω_{t_i} $(1 \leqslant i \leqslant n)$. Posons $\Omega = \bigcup_{1 \leqslant i \leqslant n} \Omega_{t_i}$ c'est un voisinage ouvert de γ dans $R \times R^{2N}$; posons :

$$\tilde{\varepsilon}^k = \min\left\{\varepsilon^k_{t_i} \ / \ 1 \leqslant i \leqslant N\right\} > 0 \qquad (1 \leqslant k \leqslant M) \ ,$$

$$\tilde{\varepsilon} = (\tilde{\varepsilon}^1, \ldots, \tilde{\varepsilon}^M)$$

Soit $(t,u) \in \Omega$ et $\varepsilon \in P(\tilde{\varepsilon})$ alors il existe i, $1 \leqslant i \leqslant u$, tel que $(t,u) \in \Omega_{t_i}$ donc $|t-t_i| < r_{t_i}$ et $|u-u_0(t_i)| < R_{t_i}$ or d'après notre façon de choisir r_{t_i}, on a $|u_0(t) - u_0(t_i)| < R_{t_i}$ d'où :

$$\|H''_{uu}(t,u,\varepsilon) - H''_{uu}(t_i,u_0(t_i),0)\| \leqslant \frac{m}{4}$$

$$\|H''_{uu}(t_i,u_0(t_i),0) - H''_{uu}(t,u_0(t),0)\| \leqslant \frac{m}{4}$$

ce qui implique, à l'aide de l'inégalité du triangle :

$$\|H''_{uu}(t,u,\varepsilon) - H''_{uu}(t,u_0(t),0)\| \leqslant \frac{m}{4} + \frac{m}{4} = \frac{m}{2}$$

en prenant δu dans R^{2N}, on peut écrire :

$$\begin{aligned} H''_{uu}(t,u,\varepsilon)(\delta u,\delta u) &= H''_{uu}(t,u_0(t),0)(\delta u,\delta u) - \\ &\quad [H''_{uu}(t,u_0(t),0)(\delta u,\delta u) - H''_{uu}(t,u,\varepsilon)(\delta u,\delta u)] \\ &\geqslant H''_{uu}(t,u_0(t),0)(\delta u,\delta u) - \\ &\quad \|H''_{uu}(t,u_0(t),0) - H''_{uu}(t,u,\varepsilon)\| \cdot |\delta u|^2 \\ &\geqslant m|\delta u|^2 - \frac{m}{2}|\delta u|^2 = \frac{m}{2}|\delta u|^2 \end{aligned}$$

ainsi, en posant $a = \frac{m}{2} > 0$, on a démontré

$$\left\{\begin{array}{l} \text{pour tout } (t,u) \in \Omega \ , \ \varepsilon \in P(\tilde{\varepsilon}) \ , \ \delta u \in R^{2N} : \\ H''_{uu}(t,u,\varepsilon)(\delta u,\delta u) \geqslant a\,|\delta u|^2 \ . \end{array}\right.$$

Comme Ω est un voisinage ouvert du compact γ , et $R \times R^{2N}$ est un espace métrique localement compact, on peut affirmer (cf. [8] p. 64) qu'il existe $\alpha > 0$ tel que $V(\gamma,\alpha) = \{(s,u) \in R \times R^{2N} /$ $\text{dist}((s,u),\gamma) < \alpha\}$ est un voisinage ouvert de γ inclus dans Ω .

L'ensemble $W(\alpha)$ de l'énoncé est clairement ouvert grâce à la continuité de l'application $(t,u) \to |u-u_0(t)|$.

Soit $(t,u) \in W(\alpha)$ et $\varepsilon \in P(\tilde{\varepsilon})$; alors il existe $s \in [0,T]$ et un entier k tels que $t = s + kT$, ainsi $u_0(t) = u_0(s)$ donc $|u-u_0(t)| < \alpha \Rightarrow |u-u_0(s)| < \alpha$. Donc $|s-s| + |u-u_0(s)| < \alpha \Rightarrow$ $\text{dist}((s,u),\gamma) < \alpha \Rightarrow (s,u) \in V(\gamma,\alpha)$, donc $(s,u) \in \Omega$, ce qui implique

$$H''_{uu}(s,u,\varepsilon)(\delta u,\delta u) \geqslant a|\delta u|^2$$

enfin de [hyp. 1] on déduit aisément : $H''_{uu}(t,u,\varepsilon) = H''_{uu}(s,u,\varepsilon)$ d'où

$$H''_{uu}(t,u,\varepsilon)(\delta u,\delta u) \geqslant a|\delta u|^2$$

■

THEOREME 2 : Sous [hyp. 1] et [hyp. 2] .

(I) il existe $\alpha > 0$, il existe $\beta > 0$, $\bar{\varepsilon} \in]0,\infty[^M$ tel que en notant $W(\alpha) = \{(t,u) \in R \times R^{2N} / |u-u_0(t)| \leqslant \alpha\}$ et

$$W'(\beta) = \{(t,v) \in R \times R^{2N} / |v-\sigma[f(t)-\dot{u}_0(t)]| < \beta\}$$

il existe une application C^r :

$I : W'(\beta) \times P(\bar{\varepsilon}) \to R^{2N}$ satisfaisant

$$\begin{cases} I(t+T,v,\varepsilon) = I(t,v,\varepsilon) \ ; \ (t,I(t,v,\varepsilon)) \in W(\alpha) \\ H'_u(t,I(t,v,\varepsilon),\varepsilon) = v \end{cases}$$

pour tout $(t,v,\varepsilon) \in W'(\beta) \times P(\bar{\varepsilon})$

(II) l'application $G : W'(\beta) \times P(\bar{\varepsilon}) \to R$ définie par $G(t,v,\varepsilon) = I(t,v,\varepsilon) \cdot v - H(t,I(t,v,\varepsilon),\varepsilon)$ est C^r et satisfait :

$$\begin{cases} G(t+T,v,\varepsilon) = G(t,v,\varepsilon) \\ G'_v(t,v,\varepsilon) = u \Rightarrow H'_u(t,u,\varepsilon) = v \end{cases}$$

pour tout $(t,v) \in W'(\beta)$, $\varepsilon \in P(\bar{\varepsilon})$.

En outre, là où existe la transformée de Legendre, sur $D(\sigma\,[f(t)-u_o(t)],\beta)$, de $u \to H(t,u,\varepsilon)$, celle-ci coïncide avec $G(t,v,\varepsilon)$.

PREUVE : Pour alléger l'écriture nous noterons $z(t) = \sigma\,[f(t)-\dot{u}_o(t)]$ Considérons la surjection canonique $\rho : R \to R/TZ = T^1$; nous savons que le tore uni-dimensionnel T^1 est une variété compacte C^∞ et ρ est un revêtement de cette variété, (cf. [4] chp. 2), en particulier autour de chaque point une double restriction de ρ est un C^∞-difféomorphisme local. Comme H est T-périodique en t, on peut définir une application $H^\bullet : T^1 \times R^{2N} \times R^M \to R$, par $H^\bullet(s,u,\varepsilon) = H(t,u,\varepsilon)$ pour $s = \rho(t) \in T^1$ et comme ρ est un C^∞-difféomorphisme local, $H^\bullet$ est C^{r+1}.

De même la T-périodicité de u_o,f,z nous permet de définir $u_o^\bullet,r^\bullet,f^\bullet : T^1 \to R^{2N}$ par $u_o^\bullet(s) = u_o(t)$, $z^\bullet(s) = z(t)$, $f^\bullet(s) = f(t)$ quand $s = \rho(t)$; et on peut appliquer le même procédé à $\dot{u}_o$, ce qui donne $(\dot{u}_o)^\bullet : T^1 \to R^{2N}$. Signalons à propos de cette dernière application qu'en définissant la dérivée, en $s \in T^1$, d'une application partant du tore comme la différentielle, en o, de cette application munie de l'accroissement constituée de l'image du nombre 1 par l'application tangente en un point t tel que $\rho(t) = s$, alors nous obtenons l'égalité entre la quotientée d'une dérivée et la dérivée d'une quotientée ainsi $\dot{\overline{u_o^\bullet}} = (\dot{u}_o)^\bullet$. Formulons l'application $X : T^1 \times R^{2N} \times R^M \times R^{2N} \to R^{2N}$

$$X(s,u,\varepsilon,v) = H_u^{\bullet\prime}(s,u,\varepsilon) - v \quad ; \quad X \text{ est } C^r .$$

Puisque u_o est solution T-périodique de (H_o) :

$$\dot{u}_o(t) = \sigma\, H_u'(t,u_o(t),0) + f(t) \quad , \text{ donc :}$$

$$\sigma\,[f(t)-\dot{u}_o(t)\,] = H_u'(t,u_o(t),0) \Rightarrow$$

$$H_u'(t,u_o(t),0) - z(t) = 0 \Rightarrow$$

$$H_u^{\bullet\prime}(\rho(t),u_o^\bullet(\rho(t)),0) - z^\bullet(\rho(t)) = 0 \Rightarrow$$

$$X(\rho(t),u_0^\bullet(\rho(t),0,z^\bullet(\rho(t)) = 0 \quad \text{pour tout } t \in R \quad \Rightarrow$$

$$X(s,u_0^\bullet(s),0,z^\bullet(s)) = 0 \quad \text{pour tout } s \in T^1 .$$

Par ailleurs, $X'_u(s,u_0^\bullet(s),0,z^\bullet(s)) = H^{\bullet\,''}_{uu}(s,u_0^\bullet(s),0)$

$= H''_{uu}(t,u_0(t),0)$ avec $\rho(t) = s$

qui est définie positive d'après [hyp. 2] donc inversible.

Ainsi, nous sommes en mesure d'appliquer le théorème des fonctions implicites et d'affirmer :

> pour tout $s \in T^1$, il existe ω_s voisinage ouvert de s dans T^1 il existe V_s voisinage de $z^\bullet(s)$ dans R^{2N}, il existe U_s voisinage ouvert de $u_0^\bullet(s)$ dans R^{2N}, il existe $\varepsilon_s^k > 0$ $(1 \leqslant k \leqslant M)$, il existe $J^s : \omega_s \times V_s \times P(\varepsilon_s) \to U_s$ application C^r telle que $X(s',J^s(s',v,\varepsilon),\varepsilon,v) = 0$ et tels que
>
> $\{(s',u,\varepsilon,v) \in \omega_s \times U_s \times P(\varepsilon_s) \times V_s \;/\; X(s',u,\varepsilon,v) = 0\} =$
>
> $\{(s',J^s(s',v,\varepsilon),\varepsilon,v) \;/\; (s',\varepsilon,v) \in \omega_s \times P(\varepsilon_s) \times V_s\}$

En outre, un $\delta > 0$ étant par avance fixé, on peut choisir ces voisinages ouverts suffisamment petits pour que

$$\omega_s \times U_s \subset \{(s,u) \in T^1 \times R^{2N} \;/\; |u-\dot{u}_0(s)| < \delta\}$$

puisque ce dernier ensemble est un voisinage ouvert de $(s,\dot{u}_0(s))$ pour tout s de T^1.

Prenons $\delta = \alpha$, où α est fourni par le théorème 1. Remarquons que les $\omega_s \times V_s$, lorsque s parcourt T^1, constituent un recouvrement ouvert du compact $K = \{(s,z^\bullet(s)) \;/\; s \in T^1\}$, donc en vertu du théorème de Heine-Borel-Lebesgue, on peut en extraire un sous-recouvrement fini : $\omega_{s_j} \times V_{s_j}$ $(1 \leqslant j \leqslant m)$.

Posons $\bar{\varepsilon}^k = \min \left(\left\{\varepsilon^k_{s_j} \ / \ 1 \leqslant j \leqslant m\right\} \cup \{\tilde{\varepsilon}^k\}\right)$ où $\tilde{\varepsilon}^k$ est fourni par le théorème 1, ainsi $\bar{\varepsilon} = (\bar{\varepsilon}^1,\ldots,\bar{\varepsilon}^M) \in]0,\infty[^M$. Nous allons maintenant montrer que nous pouvons recoller les fonctions J^{s_j}, $(1 \leqslant j \leqslant m)$, pour n'en former qu'une seule J. Notons $A = \bigcup_{1 \leqslant j \leqslant m} \omega_{s_j} \times V_{s_j}$, c'est un ouvert de $T^1 \times R^{2N}$.

Considérons $(s,v) \in A$ et $\varepsilon \in P(\bar{\varepsilon})$ et supposons que $(s,v) \in \left(\omega_{s_j} \times V_{s_j}\right) \cap \left(\omega_{s_\ell} \times V_{s_\ell}\right)$, notons $u_j = J^{s_j}(s,v,\varepsilon)$ et $u_\ell = J^{s_\ell}(s,v,\varepsilon)$ qui satisfont : $X(s,u_j,\varepsilon,v) = 0 = X(s,u_\ell,\varepsilon,v)$. Cependant l'unicité donnée par le théorème des fonctions implicites ne nous permet pas de conclure que $u_j = u_\ell$ puisqu'on ne sait pas si $u_j \in U_{s_\ell}$ ou $u_\ell \in U_{s_j}$; il nous faut donc un argument supplémentaire qui va nous être apporté par la convexité en u de H.

En effet, par notre choix des ω_s et des U_s, on a $(t,u_j),(t,u_\ell) \in W(\alpha)$ où $\rho(t) = s$; donc : $u_j,u_\ell \in D(u_o(t),\alpha) = D(\dot{u}_o(s),\alpha)$, et le théorème 1 nous permet d'affirmer que pour tout $u \in D(\dot{u}_o(s),\alpha)$, tout $\delta u \in R^{2N}$: $H''_{uu}(s,u,\varepsilon)(\delta u,\delta u) \geqslant a|\delta u|^2$.

A ce niveau,raisonnons par l'absurde, supposons $u_j \neq u_\ell$. Considérons la fonction de $[0,1]$ dans R, définie comme suit : $\lambda \to H_u'(s,(1-\lambda)u + \lambda u_j,\varepsilon)\cdot(u_j-u_\ell)$. Sa dérivée est $H''_{uu}(s,(1-\lambda)u_\ell + u_j,\varepsilon)(u_j-u_\ell,u_j-u_\ell)$ qui est strictement positive, pour tout λ dans $[0,1]$ puisque $[u_\ell,u_j] \subset D(\dot{u}_o(s),\delta)$ et avec une remarque antérieure ; ainsi la fonction considérée est strictement croissante en λ ; sa valeur en $\lambda = 1$ est strictement plus grande que sa valeur en $\lambda = 0$, donc :

$$\dot{H}_u'(s,u_j,\varepsilon)\cdot(u_j-u_\ell) > \dot{H}_u'(s,u_\ell,\varepsilon)\cdot(u_j-u_\ell) \Rightarrow$$

$$(\dot{H}_u'(s,u_j,\varepsilon) - \dot{H}_u'(s,u_\ell,\varepsilon))\cdot(u_j-u_\ell) > 0$$

or $\dot{H}_u'(s,u_j,\varepsilon) - \dot{H}_u'(s,u_\ell,\varepsilon) = v - v = 0$, c'est la contradiction cherchée, donc notre petit raisonnement par l'absurde nous a permis de conclure : $u_j = u_\ell$, soit $J^{s_j}(s,v,\varepsilon) = J^{s_\ell}(s,v,\varepsilon)$.

Cela nous permet de définir une application C^r

$$\begin{cases} J : A \times P(\bar{\varepsilon}) \to R^{2N} & \text{par} \\ J(s,v,\varepsilon) = J^{s_j}(s,v,\varepsilon) & \text{quand} \quad (s,v) \in \omega_{s_j} \times V_{s_j} \end{cases}$$

or A est un voisinage du compact K dans l'espace métrisable localement compact $T^1 \times R^{2N}$ donc il existe $\beta > 0$ tel que

$$V(K,\beta) = \{(s,v) \in T^1 \times R^{2N} \ / \ \mathrm{dist}(K;(s,v)) < \beta\} \subset A$$

Ceci nous permet de définir une application C^r

$$I : W'(\beta) \times P(\bar{\varepsilon}) \to R^{2N} \quad \text{par}$$

$$I(t,v,\varepsilon) = J(\rho(t);v,\varepsilon) \ , \quad \text{qui satisfait :}$$

$$I(t+T,v,\varepsilon) = I(t,v,\varepsilon)$$

$$H'_u(t,I(t,v,\varepsilon),\varepsilon) = v$$

$$(t,I(t),v,\varepsilon) \in W(\alpha) \qquad \text{pour tout } (t,v) \in W'(\beta) \text{ et pour tout } \varepsilon \in P(\bar{\varepsilon}) \ .$$

ce qui justifie la partie (I) de l'énoncé. Passons à la partie (II).

Au vu de sa définition, G est clairement C^r et T-périodique en t. Si $G'_v(t,v,\varepsilon) = u$, alors comme :

$$G'_v(t,v,\varepsilon)\delta v = (I'_v(t,v,\varepsilon)\delta v)\cdot v + I(t,v,\varepsilon)\cdot\delta v - H'_u(t,I(t,v,\varepsilon),\varepsilon) \cdot I'_v(t,v,\varepsilon)\delta v$$

de $H'_u(t,I(t,v,\varepsilon),\varepsilon) = v$, il vient $G'_v(t,v,\varepsilon)\delta v = I(t,v,\varepsilon)\delta v$, soit $G'_v(t,v,\varepsilon) = I(t,v,\varepsilon)$ donc $I(t,v,\varepsilon) = u \Rightarrow H'_u(t,u,\varepsilon) = v$ par définition de I.

Enfin, fixons t dans R et ε dans $P(\bar{\varepsilon})$; notons $h(u) = H(t,u,\varepsilon)$, ainsi h est une fonction numérique définie sur $D(u_o(t),\alpha)$; par un argument déjà employé dans cette preuve, du théorème 1 on déduit que si $u, u_1 \in D(u_0(t),\alpha)$ et sont distincts alors : $(h'(u)-h'(u_1))\cdot(u-u_1) > 0$, ce qui permet d'affirmer que $h' : D(u_0(t),\alpha) \to (R^N)^\star = R^N$ est injective.

Ainsi pour tout $v \in h'(D(u_0(t),\alpha))$ on peut définir la transformée de Legendre de h donnée par la formule :

$$g(v) = (h')^{-1}(v)\cdot v - h((h')^{-1}(v)) \quad , \text{ cf. [2] § 26}$$

si $v \in h'(D(u_0(t),\alpha)) \cap D(\sigma [f(t)-\dot{u}_0(t)],\beta)$ alors

$$h'(I(t,v,\varepsilon)) = v \Rightarrow I(t,v,\varepsilon) = (h')^{-1}(v)$$

$$\Rightarrow g(v) = G(t,v,\varepsilon) \qquad \blacksquare$$

<u>REMARQUE</u> : On déduit aisément du théorème précédent :

$$G''_{vv}(t,v,\varepsilon) = [H''_{uu}(t,I(t,v,\varepsilon),\varepsilon)]^{-1}$$

et bien sûr :

$$I(t,\sigma [f(t)-u_0(t)],0) = u_0(t) .$$

Enfin, comme on l'a vu au cours de la preuve $G'_v(t,v,\varepsilon) = I(t,v,\varepsilon)$, donc G'_v est C^r.

§ II : RESULTAT D'EXISTENCE

Pour tout entier k positif, on peut définir l'espace

$$C_T^k(R;R^{2N}) = \{x \in C^k(R,R^{2N}) \ / \ x \text{ est T-périodique}\}$$

c'est un espace de Banach pour la norme

$$\|x\| = \sum_{n=0}^{k} \sup \{\|x^{(n)}(t)\| \ / \ t \in [0,T]\} \quad ;$$

soit $E = \left\{w \in C_T^o(R,R^{2N}) \ / \ \int_o^T w(t)dt = 0\right\}$ c'est un espace de Banach comme sous-espace fermé de $C_T^o(R,R^{2N})$. On constate que :
$u \in C_T^1(R,R^{2N}) \Leftrightarrow \dot{u} \in E$.

L'écriture $B(w,R)$, pour $w \in E$, $R > 0$, désignera la boule ouverte de centre w et de rayon R dans E.

Introduisons deux opérateurs linéaires continus :

$$Q : E \to C_T^o(R,R^{2N}) \quad ; \quad (Qw)(t) = \int_o^t w(s)\, ds \ .$$

$$M : C_T^o(R,R^{2N}) \to E \quad ; \quad (Mx) = x - \frac{1}{T}\int_o^T x(t)\, dt \ .$$

Nous travaillons toujours sous les hypothèses (hyp. 1) et (hyp. 2) ainsi, le théorème 2 nous permet de définir l'application

$$\begin{cases} F : B(\dot{u}_o,\beta) \times P(\bar{\varepsilon}) \to E \\ F(w,\varepsilon) = M\,[Qw - G'_v(t,\sigma\,[f-w],\varepsilon)] \end{cases}$$

Un argument standard sur la différentiabilité des opérateurs de Niemytski (cf. e-g [14] § 3.9.) nous permet d'affirmer que F est C^r.

Il nous faut signaler qu'en fait notre application F est la Fréchet-dérivée, par rapport à w, de la fonctionnelle construite par CLARKE et EKELAND pour fonder le Principe Dual de Moindre Action. Ce principe est l'objet de nombreuses publications ; cf. [6], [7], [9], [10], [11], [12], [13], [3], [15],

LEMME 1 : $F(w,\varepsilon) = 0 \Rightarrow \begin{cases} \text{il existe } c \in R^{2N} \text{ tel que } u(t) = \int_0^t w(s)ds+C \\ \text{soit solution T-périodique de } (H_\varepsilon). \end{cases}$

PREUVE : Notons que $Mx = 0$ implique $x - \frac{1}{T}\int_0^T x(t)dt = 0$ donc donc $x = \frac{1}{T}\int_0^T x(t)dt$ = constante de R^{2N}. Ainsi, $F(w,\varepsilon) = 0$ entraîne : il existe $c \in R^{2N}$ tel que $Qw - G'_v(t,\sigma[f-w],\varepsilon) = -\hat{c}$; i.e.

$$Qw + \hat{c} = G'_v(t,\sigma[f-w],\varepsilon)$$

or avec le théorème 2 ceci implique :

$$H'_u(t,Qw+\hat{c},\varepsilon) = \sigma[f-w] = \sigma^{-1}[w-f]$$

ce qui se réécrit :

$$w(t) = \sigma H'_u(t,(Qw)(t)+\hat{c},\varepsilon) + f(t)$$

en posant $u(t) = \int_0^t w(s)ds + \hat{c}$, cela devient

$$\dot{u}(t) = \sigma H'_u(t,u(t),\varepsilon) + f(t)$$

et puisque $w \in E$, u est T-périodique.

LEMME 2 : Pour δw et η dans E, l'équation $F'_w(\dot{u}_o,0)\delta w = \eta$ équivaut à :

il existe $c \in R^{2N}$, tel que $y(t) = \int_0^t (\delta w)(s)ds + c$

soit solution T-périodique de l'équation, linéarisée autour de $t \to u_0(t)$, non homogène :

$$\dot{y} = \sigma H''_{uu}(t,u_0(t),0)y + \zeta(t)$$

où $\zeta(t) = - \sigma H''_{uu}(t,u_0(t),0)\eta(t)$.

<u>PREUVE</u> : Commençons par calculer la dérivée

$$F'_w(\dot{u}_0,0)\delta w = M [Q \; \delta w - G''_{uu}(t,\sigma [f-\dot{u}_0],0)\sigma^{-1} \delta w]$$

donc $F'_w(u_0,0)\delta w = \eta$ équivaut à :

il existe $c \in R^{2N}$ tel que

$$Q \; \delta w - G''_{vv}(t,\sigma [f-\dot{u}_0],0)\sigma^{-1} \delta w = \eta - c \quad \Leftrightarrow$$

$$(Q \; \delta w + c) - \eta = G''_{vv}(t,\sigma [f-\dot{u}_0],0)\sigma^{-1} \delta w$$

à l'aide de la remarque qui suit le théorème 2, ceci devient

$$(Q \; \delta w + c) - \eta = [H''_{uu}(t,I(t,\sigma \; f-\dot{u}_o \; ,0),0)]^{-1} \sigma^{-1} \delta w$$

$$= [H''_{uu}(t,u_0(t),0)]^{-1} \sigma^{-1} \delta w \quad \Leftrightarrow$$

$$\delta w = \sigma H''_{uu}(t,u_0(t),0)(Q\delta w+c) - \sigma H''_{uu}(t,u_0(t),0)\eta$$

avec les notations de l'énoncé ceci est

$$\dot{y}(t) = \sigma H''_{uu}(t,u_0(t),0)y(t) + \zeta(t) . \qquad \blacksquare$$

<u>LEMME 3</u> : Soit u une solution T-périodique de $(\mathcal{H}_\varepsilon)$ avec

$\varepsilon \in P(\bar{\varepsilon})$, $\|u-u_0\|_{C^0_T} < \delta$; $\|u-u_0\|_{C^0_T} < \beta$

(les constantes $\bar{\varepsilon},\delta,\beta$ sont celles du théorème 2) alors : $F(\dot{u},\varepsilon) = 0$

PREUVE : Puisque u est solution de $(\mathcal{H}_\varepsilon)$:

$$\dot{u}(t) = \sigma H'_u(t,u(t),\varepsilon) + f(t) \quad \text{donc}$$

$$H'_u(t,u,\varepsilon) = \sigma [f - \dot{u}]$$

l'hypothèse $\|\dot{u}-\dot{u}_0\| < \beta \Rightarrow \|\sigma [f-\dot{u}] - \sigma [f-\dot{u}_0]\| < \beta$, donc en vertu du théorème 2, $I(t,\sigma [f-\dot{u}],\varepsilon)$ est bien définie et satisfait :

$$H'_u(t,I(t,\sigma [f-\dot{u}],\varepsilon),\varepsilon) = \sigma [f-\dot{u}]$$

avec $I(t,\sigma [f(t) - \dot{u}(t)],\varepsilon) \in D(u_0(t),\delta)$. Or on a aussi par hypothèse $u(t) \in D(u_0(t),\delta)$ et comme on l'a déjà remarqué :

$u \to H'_u(t,u,\varepsilon)$ est injective sur $D(u_0(t),\delta)$, d'où

$$H'_u(t,u(t),\varepsilon) = H'_u(t,I(t,\sigma [f(t)-\dot{u}(t)],\varepsilon) \Rightarrow$$

$$I(t,\sigma [f(t)-\dot{u}(t)],\varepsilon) = u(t)$$

avec la remarque qui suit le théorème 2 :

$$G'_v(t,\sigma [f-\dot{u}],\varepsilon) = u$$

or $u = Q\dot{u} + u(0)$ donc :

$$Q\dot{u} - G'_v(t,\sigma [f-\dot{u}],\varepsilon) = -u(0) \Rightarrow$$

en appliquant M , $M(-u_0) = 0 \Rightarrow F(\dot{u},\varepsilon) = 0$. ∎

THEOREME 3 : Sous les hypothèses (hyp. 1) et (hyp. 2) si l'équation linéarisée autour de $t \to u_0(t)$

$$(L_0) \qquad \dot{y} = \sigma H''_{uu}(t,u_0(t),0)y$$

n'admet aucune solution T-périodique autre que la solution triviale, alors :

il existe $\check{\varepsilon} \in]0,\infty[^M$, il existe $\mu > 0$,
il existe une application C^r , $\varepsilon \to u_\varepsilon$ de $P(\check{\varepsilon})$ dans $C^1_T(R,R^{2N})$ telle que

. Pour $\varepsilon = 0$, u_o est la solution donnée par (hyp. 2)

. u_ε est solution T-périodique de $(\mathcal{H}_\varepsilon)$

. en outre, u_ε est la seule solution T-périodique de $(\mathcal{H}_\varepsilon)$ vérifiant $\|u_\varepsilon - u_o\|_{C^1_T} < \mu$.

<u>PREUVE</u> : Comme (L_o) n'admet aucune solution T-périodique autre que la solution nulle, alors la théorie des équations linéaires périodiques nous permet d'affirmer (cf. [18] p. 143-144, [19] p. 95) que : pour toute $b \in C^o_T(R,R^{2N})$ il existe une unique solution T-périodique de l'équation non homogène : $\dot{y} = \sigma H''_{uu}(t,u_o(t),0)y + b(t)$.

Soit $\eta \in E$, posons $\zeta(t) = - \sigma H''_{uu}(t,u_o(t),0)\eta(t)$ ainsi $\zeta \in C^o_T(R,R^{2N})$ et il existe donc une unique solution y T-périodique de $\dot{y} = \sigma H''_{uu}(t,u_o(t),0)y + \zeta(t)$ donc en vertu du lemme 2 : $F'_w(\dot{u}_o,0)\dot{y} = \eta$.

De plus, si $\delta w \in E$ satisfait $F'_w(\dot{u}_o,0)\delta w = \eta$ alors d'après le lemme 2, $t \to \int_o^t (\delta w)(s)ds + c$ est solution T-périodique de l'équation linéarisée avec ζ comme second membre, mais l'unicité implique $y(t) = \int_o^t (\delta w)(s)ds + c$, donc $\dot{y} = \delta w$. Ainsi pour tout $\eta \in E$ il existe un unique $\dot{y} \in E$ tel que $F'_w(\dot{u}_o,0)\dot{y} = \eta$; l'application $\eta \to \dot{y}$ est un inverse de $F'_w(\dot{u}_o,0)$, avec le théorème de Banach cet inverse est automatiquement continu, donc $F'_w(\dot{u}_o,0)$ est un automorphisme topologique de E.

Le lemme 3 nous donne $F(\dot{u}_o,0) = 0$; nous pouvons alors appliquer le théorème des fonctions implicites qui nous assure l'existence d'un $\varepsilon' \in]0,\infty[^M$, d'un $R > 0$ $(R \leqslant \beta)$ et d'une application $\varepsilon \to w_\varepsilon$ C^r de $P(\varepsilon')$ dans $B(\dot{u}_o,R)$ tels que $w_o = \dot{u}_o$, $F(w_\varepsilon,\varepsilon) = 0$ pour tout ε dans $P(\varepsilon')$ et en outre

$$\{(w,\varepsilon) \in B(u_o,R) \times P(\varepsilon') \ / \ F(w,\varepsilon) = 0\} =$$

$$= \ \{(w_\varepsilon,\varepsilon) \ / \ \varepsilon \in P(\varepsilon')\} \ .$$

D'après le lemme 1, il existe $c_\varepsilon \in R^{2N}$ tel que $u_\varepsilon(t) = \int_o^t w_\varepsilon(s)ds + c_\varepsilon$ soit solution T-périodique de (H_ε) . Si l'on considère l'opérateur Q de E dans $C^1_T(R,R^{2N})$, il est linéaire continu donc C^∞. On remarque dans la preuve du lemme 1 que $c_\varepsilon = - Qw_\varepsilon + G'_v(t,\sigma\,[f-w_\varepsilon],\varepsilon)$, (ici l'opérateur Q est considéré de E dans $C^o_T(R,R^{2N})$), or l'application $(w,\varepsilon) \to -Qw + G'_v(t,\sigma\,[f-w\,],\varepsilon)$ est C^r de $B(\dot{u}_o,R) \times P(\varepsilon')$ dans $C^o_T(R,R^{2N})$; ainsi en composant des applications C^r : $\varepsilon \to c_\varepsilon$ est C^r de $P(\varepsilon')$ dans $C^o_T(R,R^{2N})$, mais c_ε est constante et $in : R^{2N} \to C^1_T(R,R^{2N})$ désignant l'injection insertion est clairement linéaire isométrique danc C^∞, ainsi on a

$$u_\varepsilon = Qw_\varepsilon + in(c_\varepsilon) \text{ et } \varepsilon \to u_\varepsilon \text{ est } C^r \text{ de } P(\varepsilon') \text{ dans } C^1_T(R,R^{2N})$$

comme somme de deux applications C^r.

Posons $\mu = \min\,\{\alpha,R\} > 0$; la continuité de l'application $\varepsilon \to u_\varepsilon$ nous permet d'affirmer qu'il existe $\check{\varepsilon} \in R^M$, $0 < \varepsilon^k \leq \min\,\{\varepsilon'^k,\bar{\varepsilon}^k\}$; tel que $\|u_\varepsilon-u_o\|_{C^1_T} < \mu$ dès que $\varepsilon \in P(\check{\varepsilon})$.

Considérons $\varepsilon \in P(\check{\varepsilon})$ et u solution T-périodique de (H_ε) vérifiant $\|u-u_o\|_{C^1_T} < \mu$; alors par définition de la norme C^1_T on a $\|\dot{u}-\dot{u}_o\|_E < \mu$, donc avec le lemme 3 on obtient $F(\dot{u},\varepsilon) = 0$ puisque $\mu \leq R \leq \beta$ et $\mu \leq \alpha$; de plus l'unicité donnée par le théorème des fonctions implicites induit : $\dot{u} = w_\varepsilon$; de là, il vient $H'_u(t,u(t),\varepsilon) = H'_u(t,u_\varepsilon(t),\varepsilon)$. Or, on a $u(t),u_\varepsilon(t) \in D(u_o(t),\alpha)$, car $\mu \quad \alpha$, et comme on l'a déjà vu à plusieurs reprises l'application $u \to H'_u(t,u,\varepsilon)$ est injective sur $D(u_o(t),\alpha)$, ce qui implique $u(t) = u_\varepsilon(t)$.

∎

§ III : OSCILLATEUR ANHARMONIQUE

Considérons les équations vectorielles du second ordre

$$(S_\varepsilon) \qquad \ddot{q} + \Omega_q + \varepsilon V'_q(t,q) = e(t)$$

où

$$\begin{cases} q \in R^N \; ; \; \Omega \text{ est une matrice } N \times N \; ; \\ e : R \to R^N \text{ est continue et T-périodique} \\ V : R \times R^N \to R \text{ est } C^{r+1} \quad (r \geqslant 2) \quad \text{et} \\ V(t+T,q) = V(t,q) \quad \text{pour tous } t,q \, . \end{cases}$$

en définissant le hamiltonien $H(t,q,p,\varepsilon) = \frac{1}{2}\,\Omega q \cdot q + \frac{1}{2}\,p \cdot p + \varepsilon V(t,q)$ alors le système hamiltonien perturbé :

$$\begin{cases} \dot{q} = p \\ \dot{p} = -\,\Omega q - \varepsilon\, V'_q(t,q) + e(t) \end{cases}$$

est exactement l'écriture en équation du premier ordre de (S_ε).

<u>THEOREME 4</u> : Supposons Ω définie positive ; supposons que l'équation linéaire homogène $\ddot{q} + \Omega q = 0$ n'ait aucune autre solution T-périodique que la solution triviale. Alors il existe $\bar{\varepsilon} > 0$, $\theta > 0$, il existe une application $\varepsilon \to q_\varepsilon$, C^r de $]-\bar{\varepsilon},\bar{\varepsilon}\,[$ dans $C^2_T(R,R^N)$, telle que q_ε soit solution T-périodique de (S_ε) et de plus, soit la seule solution T-périodique de (S_ε) à satisfaire $\|q_\varepsilon - q_o\|_{C^2_T} < \theta$.

PREUVE : Nous allons utiliser le théorème 3. L'hypothèse (hyp. 1) est trivialement vérifiée. Nous avons supposé que $\ddot{q} + \Omega q = 0$ n'admet aucune solution T-périodique, alors d'après l'alternative de Fredholm : $\ddot{q} + \Omega q = e(t)$ admet une unique solution T-périodique que nous noterons q_o . Pour revenir au formalisme du théorème 3, on a $u = (q,p)$; on posera donc $u_o = (q_o,\dot{q}_o)$, ainsi $H''_{uu}(t,(q_o,\dot{q}_o),0) = \begin{vmatrix} \Omega & 0 \\ 0 & id \end{vmatrix}$ qui est bien définie positive. L'hypothèse (hyp. 2) est satisfaite. Nous pouvons alors utiliser les conclusions du théorème 3, ce qui nous donne l'existence d'un $\bar{\varepsilon}>0$, d'un $\mu>0$, et d'une application $\varepsilon \to u_\varepsilon = (q,\dot{q}_\varepsilon)$ C^r de $]-\bar{\varepsilon},\bar{\varepsilon}[$ dans $C^1_T(R,R^N \times R^N)$; mais puisque $\|(q,\dot{q})\|_{C^1_T(R,R^{2N})} \geqslant \|q\|_{C^2_T(R,R^N)}$ on en déduit que $\varepsilon \to q_\varepsilon$ est C^1 de $]-\varepsilon,\varepsilon[$ dans $C^2_T(R,R^N)$; et en ce qui concerne l'unicité, il suffira de prendre $\theta = \mu$. ■

REMARQUE : Comme Ω est définie positive, alors (cf. [1] p. 491-492 ou [2] p. 108) l'équation $\ddot{q} + \Omega q = 0$ peut se ramener à N oscillateurs non couplés $\ddot{q}^k + \lambda_k^2 q^k = 0$ (k=1,...,N), c'est-à-dire que le champ hamiltonien de cette équation peut s'écrire sous la Forme Normale Réelle. Dans ce cas si nous faisons l'hypothèse de non récurrence, i.e. les $\lambda_1,\ldots,\lambda_N$ sont deux à deux indépendants sur Z, alors les seules périodes minimales qui puissent correspondre à des solutions périodiques de $\ddot{q} + \Omega q = 0$ sont $\frac{2\pi}{\lambda_k}$; donc si nous prenons $T < \frac{2\pi}{\lambda_k}$ (pour tout k) où $T \notin \frac{2\pi}{\lambda_k} Z$ (pour tout k) alors nous pourrons affirmer que $\ddot{q} + \Omega q = 0$ n'a aucune solution T-périodique.

§ IV : QUAND L'EXCITATION SERT DE PERTURBATION

Considérons pour $f \in C^o_T(R,R^{2N})$ l'équation

$$(\mathcal{H}_f) \qquad \dot{u} = \sigma H'_u(t,u) + f(t)$$

où

$$\begin{cases} H : R \times R^{2N} \to R \quad \text{est } C^{r+1} \quad (r \geqslant 2) \\ H(t+T,u) = H(t,u) \quad \text{pour tous } t,u \end{cases}$$

Nous ferons dans tout ce paragraphe l'hypothèse suivante :

(hyp. 3) $\begin{cases} \text{on suppose comme une solution T-périodique } u_o \text{ de } (\mathcal{H}_{f_o}) \\ \text{et } H'_{uu}(t,u_o(t)) \text{ est définie positive, pour tout } t \end{cases}$

<u>LEMME 4</u> : Il existe $\beta' > 0$ tel que l'on puisse définir l'application

$$A : B_E(\dot{u}_o,\beta') \times B_{C^o_T}(f_o,\beta') \to E$$

$$A(w,f) = M\,[Qw - G'_v(t,\sigma\,[f-w\,])]$$

<u>PREUVE</u> : Ceci découle du théorème 2 ; en effet nous sommes ici dans le cas indépendant de ε , ainsi on sait définir :

$$I : W'(\beta) \to R^{2N} \quad \text{et} \quad G : W'(\beta) \to R$$

en posant $\beta' = \frac{1}{2}\beta$; si $\|f-f_o\|_{C^o_T} < \beta'$ et $\|w-\dot{u}_o\| < \beta'$, alors

$|\sigma(f(t)-w(t)) - \sigma(f_o(t)-\dot{u}_o(t))| < \beta' + \beta' = \beta$ donc
$(t,\sigma(f(t)-w(t))) \in W'(\beta)$ donc $G'_v(t,\sigma(f(t)-w(t)))$ est définie donc
$A(w,f)$ est définie. ■

<u>LEMME 5</u> : $A(w,f) = 0 \Rightarrow$ $\begin{cases} \text{il existe } c \in R^{2N} \text{ tel que } u(t) = \int_o^t w(s)ds+c \\ \text{soit solution T-périodique de } (H_f). \end{cases}$

<u>PREUVE</u> : $A(w,f) = 0 \Rightarrow$ il existe c constante de R^{2N} telle que

$$Qw - G_v'(t,\sigma[f-w]) = -c \quad , \quad \text{i.e.}$$

$$Qw + c = G_v'(t,\sigma[f-w]) \quad , \quad \text{et avec le théorème 2}$$

$$H_u'(t,Qw+c) = \sigma^{-1}[w-f] \quad , \quad \text{soit}$$

$$w(t) = \sigma H_u'(t,(Qw)(t)+c) + f(t) \quad , \quad \text{c'est-à-dire}$$

$$\dot{u}(t) = \sigma H_u'(t,u(t)) + f(t) \quad .$$

■

<u>LEMME 6</u> : Soit u une solution T-périodique de (H_f) avec $\|f-f_o\| < \beta'$, $\|\dot{u}-\dot{u}_o\| < \beta'$, $\|u-u_o\| < \alpha$; alors $A(\dot{u},f) = 0$.

<u>PREUVE</u> : Comme $\dot{u}(t) = \sigma H_u'(t,u(t)) + f(t)$ on obtient $H_u'(t,u) = \sigma[f-\dot{u}]$ Par ailleurs $\|\dot{u}-\dot{u}_o\| < \beta'$ et $\|f-f_o\| < \beta' \Rightarrow \|\sigma[f-\dot{u}] - \sigma[f_o-\dot{u}_o]\| < \beta$, donc $I(t,\sigma[f-\dot{u}])$ est défini et satisfait

$$H_u'(t,I(t,\sigma[f-\dot{u}]) = \sigma[f-\dot{u}]$$

avec $I(t,\sigma[f-\dot{u}](t)) \in D(u_o(t),\alpha)$ et l'injectivité de $u \to H_u'(t,u)$ sur $D(u_o(t),\alpha)$ nous permet d'affirmer

$$I(t,\sigma[f-\dot{u}]) = u \quad ; \quad \text{i.e.} \quad G_v'(t,\sigma[f-\dot{u}]) = u = Q\dot{u} + u(0)$$

ainsi $Q\dot{u} - G_v'(t,\sigma[f-\dot{u}]) = \text{constante} \Rightarrow A(\dot{u},f) = 0$.

■

<u>LEMME 7</u> : Pour $\delta w,\eta$ dans E, l'équation $A_w'(\dot{u}_o,f_o)\delta w = \eta$ équivaut à : il existe $c \in R^{2N}$ tel que $y(t) = \int_o^t (\delta w)(s)ds + c$ soit solution T-périodique de l'équation, linéarisée autour de u_o,

non homogène : $\dot{y} = \sigma H''_{uu}(t,u_o(t))y + \zeta(t)$

où $\zeta(t) = -\sigma H''_{uu}(t,u_o(t))\eta(t)$.

PREUVE : Très similaire à celle du lemme 2. ∎

THEOREME 5 : Sous (hyp. 3), et sous la condition additionnelle :

l'équation linéarisée autour de u_o :

(L_{f_o}) $\dot{y} = \sigma H''_{uu}(t,u_o(t))y$

n'admet aucune solution T-périodique non-nulle

alors : il existe $\bar{\beta} > 0$, il existe $\nu > 0$, il existe une application C^r $f \to u_f$ de $B_{C^o_T}(f_o,\bar{\beta})$ dans $C^1_T(R,R^{2N})$

telle que :
- u_f soit solution de (H_f)
- $u_o = u_{f_o}$
- u_f soit la seule solution de (H_f) à satisfaire $\|u_f - u_o\|_{C^1_T} < \nu$.

PREUVE : La condition sur (L_{f_o}), le lemme 7, et l'alternative de Fredholm nous permettent de dire que $A'_w(\dot{u}_o,f_o)$ est un automorphisme topologique de E ; par le lemme 6 on sait que $A(\dot{u}_o,f_o) = 0$; ainsi nous pouvons appliquer le théorème des fonctions implicites et exhiber une application C^r $f \to w_f$ d'une boule de $C^o_T(R,R^{2N})$ dans E. La conclusion est similaire à celle de la preuve du théorème 3. ∎

§ V : UN CAS DE BIFURCATION

Considérons un hamiltonien de la forme

$$H(t,u,\varepsilon) = H^{o}(t,u) + \varepsilon H^{1}(t,u) \quad \text{où} \quad u \in R^{N} \times R^{N}, \ \varepsilon \in R$$

$$\begin{cases} H^{o}(t+t,u) = H^{o}(t,u), \quad H^{1}(t+T,u) = H^{1}(t,u) \\ H^{o} \text{ et } H^{1} \quad \text{sont} \quad C^{r+1} \quad (r \geqslant 2) \end{cases}$$

On considère les équations hamiltoniennes

$$(H_{\varepsilon}) \qquad \dot{u} = \sigma H'_{u}(t,u,\varepsilon) \quad .$$

<u>THEOREME 6</u> : On suppose que

$$\begin{cases} \cdot \ H^{o\prime}_{u}(t,0) = 0 = H^{1\prime}_{u}(t,0) \\ \cdot \ H^{o\prime\prime}_{uu}(t,0) \text{ et } H^{1\prime\prime}_{uu}(t,0) \quad \text{sont définies positives} \end{cases}$$

Le système linéarisé (L_o) $\dot{y} = \sigma H^{o\prime\prime}_{uu}(t,0)y$ a une seule (à un facteur scalaire près) solution T-périodique non-triviale.

<u>Alors</u> : Il existe $\hat{\varepsilon} > 0$, $\rho > 0$, il existe une application C^{r} , $\varepsilon \to u_{\varepsilon}$ de $]-\hat{\varepsilon},\hat{\varepsilon}[$ dans $C^{1}_{T}(R,R^{2N})$ telle que

$u_{o} = 0$, et pour $\varepsilon \neq 0$: u est solution T-périodique non-nulle de (H_{ε}). De plus les seules solutions T-périodiques de (H_{ε}) qui sont dans $B_{C^{1}_{T}}(0,\rho)$ sont exactement 0 et u_{ε} .

PRÉUVE : Nous pouvons utiliser les résultats du § I avec $f = 0$; ainsi nous avons $F : B_E(0,\beta) \times]-\bar{\varepsilon},\bar{\varepsilon}[\to E$ de classe C^r

$$F(w,\varepsilon) = M\,[Qw-G'_v(t,-\sigma w,\varepsilon)] = M\,[Qw-I(t,-\sigma w,\varepsilon)]$$

Compte-tenu de l'hypothèse sur les gradients et de l'injectivité de $u \to H'_u(t,u,\varepsilon)$: $F(0,\varepsilon) = 0$ pour tout ε . Donc en dérivant par rapport à ε :

(i) $$F'_\varepsilon(0,0) = 0$$

Par ailleurs compte-tenu de l'hypothèse sur (L_0) et du lemme 2, on a :

(ii) $\ker F'_w(0,0)$ est un s.e.v. de dimension 1

Nous désignerons par x_0 un générateur de ce sous-espace et avec l'alternative de Fredholm :

(iii) $\operatorname{im} F'_w(0,0) = Y_1$ est un s.e.v. de codimension 1 .

De l'égalité $H'_u(t,I(t,v,\varepsilon),\varepsilon) = v$, il vient en dérivant par rapport à ε :

$$H''_{uu}(t,I(t,v,\varepsilon),\varepsilon)\cdot I'_\varepsilon(t,v,\varepsilon) + H''_{u\varepsilon}(t,I(t,v,\varepsilon),\varepsilon) = 0$$

dérivons encore par rapport à $\varepsilon \Rightarrow$

$$H''_{uuu}(t,I(t,v,\varepsilon),\varepsilon)(I'_\varepsilon(t,v,\varepsilon),I'_\varepsilon(t,v,\varepsilon)) + H''_{uu\varepsilon}\cdot I'_\varepsilon(t,v,\varepsilon) +$$
$$H''_{uu}(t,I(t,v,\varepsilon),\varepsilon)I''_{\varepsilon\varepsilon}(t,v,\varepsilon) + H'''_{uu\varepsilon}(t,I(t,v,\varepsilon),\varepsilon)I'_\varepsilon(t,v,\varepsilon) +$$
$$H'''_{u\varepsilon\varepsilon}(t,I(t,v,\varepsilon),\varepsilon) = 0 \quad ;$$

faisons $\varepsilon = 0$, $v = 0$ alors $I'_\varepsilon(t,0,0) = 0$ et $H''_{u\varepsilon\varepsilon} = 0$, ce qui donne

$$H''_{uu}(t,0,0)\cdot I''_{\varepsilon\varepsilon}(t,0,0) = 0 \quad \text{or} \quad H''_{uu}(t,0,0) = H^{0}{}''_{uu}(t,0)$$

qui est inversible donc nécessairement $I''_{\varepsilon\varepsilon}(t,0,0) = 0$ or $F''_{\varepsilon\varepsilon}(0,0) = M\,[-\,I''_{\varepsilon\varepsilon}(t,0,0)]$ donc :

(iv) $$F''_{\varepsilon\varepsilon}(0,0) = 0$$

. On calcule aisément que $I''_{v\varepsilon} = - H''_{uu}(t,I,\varepsilon)^{-1} H'''_{uu} H''_{uu}(t,I,\varepsilon)^{-1}$ ainsi $F'_{w\varepsilon}(0,0)x_o = M\,[I''_{v\varepsilon}(t,0,0)\,\sigma\, x_o(t)]$.

. Comme $x_o \in \ker F'_w(0,0)$, $x_o \neq 0$, alors $x_o = \dot{y}_o$ où y_o est solution T-périodique de (L_o) ; ainsi $x_o = \dot{y}_o = \sigma\, H^{o''}_{uu}(t,0)y_o$. L'équation adjointe de (L_o) est

$$(L_o^{\star}) \qquad \dot{z} = H^{o''}_{uu}(t,0)\,\sigma\, z$$

donc $\sigma^{-1} y_o$ est solution T-périodique de $(L_o^{\star})$: c'est même la seule à un facteur scalaire près.

L'alternative de Fredholm nous dit que $\eta \in Y_1$ si et ssi

$$\int_o^T (-\sigma\, H^{o''}_{uu}\,\eta)\cdot(\sigma^{-1} y_o)dt = 0 \iff \int_o^T (H^{o''}_{uu}\cdot\eta)\; y_o\; dt = 0$$

$$\iff \int_o^T \eta\cdot(H^{o''}_{uu}\cdot y_o)dt = 0$$

$$\iff \int_o^T \eta\cdot\sigma x_o\; dt = 0$$

Donc l'assertion $F'_w(0,0)x_o \notin Y_1$ équivaut à l'assertion

$$\int_o^T (F'_w(0,0)x_o \cdot \sigma x_o)\; dt \neq 0$$

Nous allons calculer cette intégrale, elle vaut :

$$\int_o^T - [H^{o''}_{uu}]^{-1} \cdot H^{1''}_{uu} \cdot [H^{o''}]^{-1}\; \sigma\, x_o(t)\cdot\sigma x_o(t)$$

$$- \frac{1}{T} \left(\int_o^T (F''_w\, x_o)dt\right) \cdot \underbrace{\left(\int_o^T \sigma\, x_o\; dt\right)}_{= 0} \qquad \text{car } x_o \in E$$

$$= \int_0^T H^{1}_{uu}{}''(t,0)\,[H^{o}_{uu}{}'']^{-1}\,[H^{o}_{uu}{}'']\,y_o(t)\cdot [H^{o}_{uu}{}'']^{-1}\,[H^{o}_{uu}{}'']\,y_o(t)\,dt$$

$$= \int_0^T H^{1}_{uu}{}''(t,0)y_o(t)\cdot y_o(t)\,dt$$

or $H^{1}_{uu}{}''(t,0)$ défini positif pour tout t et T-périodique en t. Il existe donc $m > 0$ tel que

$$\int_0^T H^{1}_{uu}{}''(t,0)y_o(t)\cdot y_o(t)dt \geqslant m \int_0^T |y_o(t)|^2\,dt > 0$$

car y_o est continue et non-nulle.

Ainsi $\int_0^T F''_{w\varepsilon}(0,0)x_o(t)\cdot \sigma x_o(t)\,dt > 0 \Rightarrow$

(v) $F''_{w\varepsilon}(0,0)x_o \notin Y_1$

Avec (i), (ii), (iii), (iv), (v) les hypothèses du théorème de Crandall-Rabinowitz sont satisfaites (cf. [16] p. 77) donc (0,0) est point de bifurcation de F ; ainsi $\varepsilon \to (0,\varepsilon)$ est une branche et $\varepsilon \to (\varepsilon,w_\varepsilon)$ une autre branche qui ne se coupent qu'en (0,0) ; $w_\varepsilon = \dot{u}_\varepsilon$ où u_ε est solution T-périodique de (H_ε) ; on remonte à $\varepsilon \to u_\varepsilon$ comme dans la preuve du théorème 3.

∎

BIBLIOGRAPHIE

[1] ABRAHAM (Ralph) et MARDSEN (Jerrold E.) : "Foundations of Mechanics" second edition. Benjamin 1980.

[2] ARNOLD (Vladimir I.) : "Méthodes mathématiques de la mécanique classique". Mir. 1976.

[3] AZE (Dominique) et BLOT (Joël) : "Systèmes hamiltoniens : leurs solutions périodiques". CEDIC (à paraître).

[4] BERGER (Marcel) et GOSTIAUX (Bernard) : "Géométrie différentielle" Coll. U. Armand Colin. 1972.

[5] BLOT (Joël) : "Invariance en Calcul des Variations et méthodes non-linéaires en dynamique hamiltonienne". Thèse de 3e cycle. 1981. CEREMADE. Paris IX-Dauphine 75775-Paris Cedex 16.

[6] CLARKE (Franck H.) et EKELAND (Ivar) : "Solutions périodiques, de période donnée, des équations hamiltoniennes". CRAS. t. 287. (27 nov. 1978) série A. 1013.

[7] CLARKE (Franck H.) et EKELAND (Ivar) : "Non-linear oscillations and boundary-value problems for hamiltonian systems". J. Rat. Mech. Ana. (à paraître).

[8] DIEUDONNE (Jean) : "Fondements de l'analyse moderne ; Eléments d'analyse, tome 1". Gauthier-Villars. 1969.

[9] EKELAND (Ivar) : "Periodic solutions of hamiltonian equations and a theorem of P. Rabinowitz". J. of Diff. Equa. 34 (523-534) 1979.

[10] EKELAND (Ivar) : "Forced oscillations of non-linear hamiltonian systems, II". Advances in math. (à paraître).

[11] EKELAND (Ivar) : "Oscillations de systèmes hamiltoniens non-linéaires,III". Bull. Soc. Math. France 109. 1981. (297-330).

[12] EKELAND (Ivar) : "A perturbation theory near convex hamiltonian systems". Technical Report n° 82-1. January 1982. University of British-Columbia-Vancouver- Canada.

[13] EKELAND (Ivar) et LASRY (Jean-Michel) : "On the number of periodic trajectories for a hamiltonian flow on a convex energy surface". Annals of Math. 112. 1980. (283-319).

[14] FLETT (T.M.) : "Differential analysis". Cambridge University Press. 1980.

[15] GAUSSENS (Erick) : "Quelques propriétés topologiques de l'ensemble des solutions périodiques d'un système différentiel non-linéaire". Cahier du CEREMADE N° 8201. 1982.

[16] NIRENBERG (Louis) : "Topics in non-linear functional analysis" Courant Institute of Mathematical Sciences. New-York University. 1974.

[17] ROCKAFELLAR (R. Tyrell) : "Convex analysis". Princeton University Press. 1970.

[18] ROUCHE (Nicolas) et MAWHIN (Jean) : "Equations différentielles ordinaires : tome 1 : théorie générale". Masson. 1973.

[19] ROUCHE (Nicolas) et MAWHIN (Jean) : "Equations différentielles ordinaires : tome 2 : stabilité et solutions périodiques". Masson 1973.

"NUMERICAL RESEARCH OF PERIODIC SOLUTION FOR A HAMILTONIAN SYSTEM"

E. GAUSSENS

CEREMADE
I.P.S.N.

§ 0. INTRODUCTION

This paper deals with the following problem : Find approximate solutions of

$$(\mathcal{H}) \qquad \begin{cases} \dfrac{dy}{dt} = \dfrac{\partial H}{\partial q}(y,q) + f_1(t) \\ \\ \dfrac{dq}{dt} = -\dfrac{\partial H}{\partial y}(y,q) + f_2(t) \end{cases} \qquad t \in [0,T]$$

with T pre-assigned, and the boundary constraint :

$$[B.C] \qquad \begin{cases} y(0) = y(T) \in R^n \\ \\ q(0) = q(T) \in R^n \end{cases}$$

We recognize in $(\mathcal{H})$ a Hamiltonian system with forcing terms f_1 and f_2. Finding periodic solutions of $(\mathcal{H})$ is an old and classical problem in mechanics. It is well known [cf. [9], [10]] that true solutions of $[\mathcal{H}]$ and [B.C] are related, by the "Principle of Maupertuis" and the Euler-Lagrange equations, to a problem in Calculus of variation involving the so-called "Action Integral".

But existence theorems and numerical methods are not very easy to construct from these results [see numerical applications in [11] forexample].

A. Bahri and H. Beresticky [cf [8]] proved a global result of existence using partial-differential equation techniques. Although this result is very complete, it seems to have two drawbacks : one of their hypothesis is not so easy to verify, and their proof is not constructive at all and so numerical methods are not immediately available.

F. Clarke and I. Ekeland [cf [1] → [4]] introduced another problem in the calculus of variation based on their "dual action integral" and related to our problem by their "dual action principle" which acts like the "Principle of Maupertuis".

From these results, they have proved different kind of existence theorem (the more general is in Ekeland [4]). Those theorems are the basis of this work. But they also gave other results : superquadratic potential [6] perturbation theory [13]. In [15] one may find an extensive description of [1] to [4] together with the classical approach.

In [14] D. Azé showed that the solutions "A la Ekeland" coexist with another kind of solutions introduced by Rabinowitz, Ambrosetti-Rabinowitz in [5], [6] and [7]. In [17] one may find some topological properties of the set of solutions (y,q,f) of (H) and [B.C] using the Ekeland approach (as a consequence one may show that the solution "A la Ekeland" and "A la Rabinowitz" are connected by a continuous path of element of this set).

§ 1. Formulation and assumption.

1.1. The general problem.

Let J be the Clarke-Ekeland Integral defined below ; under the assumption that H is convex

$$J : L^2(0,T;\mathbb{R}^n) \times L^2(0,T;\mathbb{R}^n) \times L^2(0,T,\mathbb{R}^n)^2 \to \mathbb{R}^+ \cup \{+\infty\}$$

$$(x,p,f) \to J(x,p,f)$$

with

$$J(x,p,f) = D(x,p) + \int_0^T G(-p(t)+f_2(t),x(t)-f_1(t))\,dt$$

where

$$D(x,p) = \frac{1}{2}\int_0^T \left\{p(t)\int_0^t x(s)ds - x(t)\int_0^t p(s)ds\right\} dt$$

(G represents the Fenchel conjuguate of H [see [20], [21]]).

The set issued from the constraint [B.C] is :

$$E = \left\{(x,p) \in L^2(0,T;\mathbb{R}^n)^2 \ / \int_0^T x(t)dt = \int_0^T p(t)dt = 0\right\}$$

The "dual action principle" then states that the two following proposition are equivalent, where J'( ) denotes the derivative of J relative to (x,p) :

(1) $$P_E(J'(x,p,f)) = 0$$

(2) $\exists\,(c_1,c_2) \in \mathbb{R}^{2n}$ such that the two functions :

$$y(\cdot) = \int_0^{\cdot} x\,dt + c_1$$

$$q(\cdot) = \int_0^{\cdot} p\,dt + c_2$$

are solutions of $(\mathcal{H})$ and [B.C] (with the same f, of course)

(when P_E denotes the projection of $(L^2(0,T,\mathbb{R}^n))^2$ on E) .

To prove the existence result, it is shown that under appropriate assumption one may find solutions to that problem :

$$(\mathcal{P}_E) \qquad \left\{ \operatorname*{Inf}_{(x,p)\in E} J(x,p,f) \right\}$$

and so points verifying (1).

Here are the assumptions, which, from now on, we shall refer as $[H_o]$ (see [4] for more detail) :

(1) $H : \mathbb{R}^{2n} \to \mathbb{R}^+$ convex and C^2

(2) $H(a) = 0 \Leftrightarrow a = 0$

(3) $r^{-1}\,\mathrm{Min}\,\{H(u) \,/\, |u| = r\} \to +\infty$ as $r \to +\infty$

(4) $G(m_2,-m_1) > \operatorname*{Inf}_{s>0} \left\{\phi(s+\|F\|_\infty) + |m|(s+\|F\|_\infty) - \frac{\pi}{T} s^2\right\}$

where

(•) F is the primitive of f such that $\int_0^T F(t)\,dt = 0$

(•) ϕ is defined for $s > 0$ by :

$$\phi(s) = \mathrm{Max}\,\{H(u) \,/\, |u| = s\}$$

(•) $m = (m_1, m_2)$ is defined by

$$m = \frac{1}{T}\int_0^T f(t)\,dt$$

<u>Example</u>

Suppose $H(a,b) = \frac{1}{2} b^2 + V^\star(a)$

where

(1) $V^\star$ is convex and C^2

(2) $V^\star(a) > V^\star(0) = 0 \quad \forall\, a \neq 0$

(3) $r^{-1}V^\star(r) \to +\infty$ as $r \to +\infty$

(4) $\exists\, k,\beta \,/\, k > 0\,,\ \beta > 2$

$$V(a) < \frac{k^\beta}{\beta}\,|a|^\beta \qquad \forall\, a \in \mathbb{R}^n$$

Then A satisfies $[H_o]$ ($V^\star$ denotes the conjugate of V).

Remark

Under $[H^o]$ G is simply the Legendre transform of H [cf. [21]]. In our last example we find :

$$V(a) = \frac{3}{4}\,|a|^{4/3}$$

if

$$V^\star(a) = \frac{1}{4}\,|a|^4$$

1.2. How to reduce (P_E) to a finite dimensionnal problem ?

A natural way to reduce (P_E), as long as we are interested by periodic solutions is to use the Fourier expansion together with a Galerkin method. We tried this way with little success. In particular one of the obstacles is the difficulty to know how to stop in the expansion considering that all the approximate solutions in this way are exactly periodic. This problem is not easy to evade, one could be convinced of that by reading the Fermi-Pasta-Ulam. paper [11].

Here, we choose to discretize the whole line of time and consider "step" functions.

We are then able to establish a discretised version of the "dual action principle" and this gives us a way to appreciate the accuracy of our approximate solutions. [The choice of step functions instead of piece wise linear one for instance can be justified by the fact that x and p are (perhaps distributionnaly) derivatives of y and q].

We shall denote, from now on, as v the couple (y,q), and introduce the relation matrix :

$$\sigma \ = \ \left. \begin{array}{c|c} 0 & I_n \\ \hline -I_n & 0 \end{array} \right| \begin{array}{l} \}\ n \\ \\ \}\ n \end{array}$$
$$\underbrace{\qquad}_{n}\ \underbrace{\qquad}_{n}$$

So we can write J and E as

$$J(v) \ = \ \int_o^T \left\{ \frac{1}{2} \left[\sigma v(t) \ , \ \int_o^t v(\tau) d\tau \right] + G(-\sigma v(t) + f(t)) \right\} dt$$

$$E \ = \ \left\{ v \in L^2(0,T,\mathbb{R}^{2n}) \ / \ \int_o^T v(\tau) \ d\tau \ = \ 0 \right\}$$

We shall partition the real line in this way :

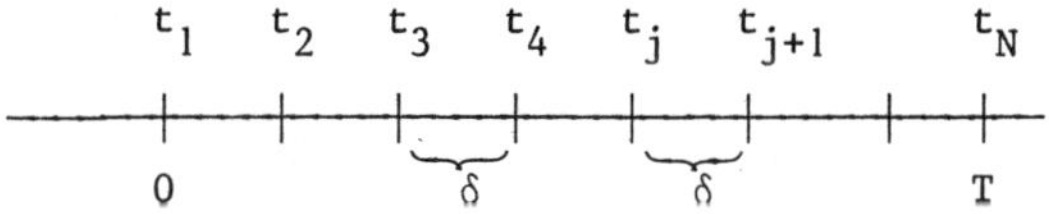

with $\delta = T/(N-1)$

and consider step functions:

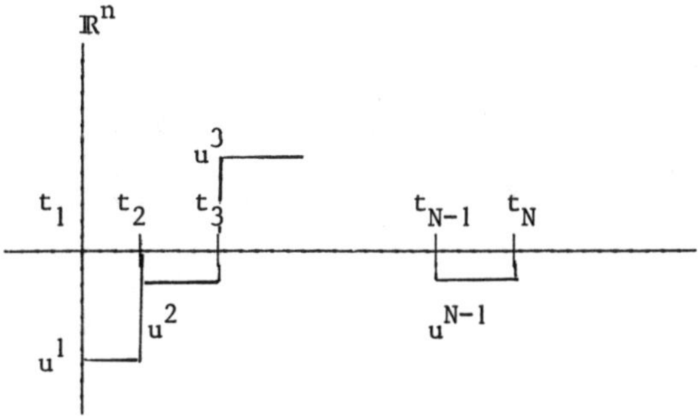

We shall, then, denote by v (= (v_1, v_2)), an element of $(\mathbb{R}^{np})^2$ where p is equal to N-1. If one wishes so v_i might be seen as a matrix of n rows and p lines where each row corresponds to the value of the step function at the point t_j , $j = 1$, p , and v^j is the value of v at t_j.

In that case J and E can be rewritten in the space of step functions, and we shall denote them by J_N and E_N, although they are the same function and space. We have :

$$E_N = \left\{ u \in (\mathbb{R}^{np})^2 \;/\; \sum_{j=1}^{p} u_i^j = 0 \quad i=1,2 \right\}$$

$$J_N(v) = \frac{1}{2}\,\delta^2 \sum_{j=1}^{p-1} s_j(v) + \delta \sum_{j=1}^{p} g_j(v)$$

where

$$\delta = T/p = T/N-1$$

$$s_j(v) = \sum_{k=1}^{j} \left(u_2^j\, u_1^k - u_1^j\, u_2^k\right)$$

$$g_j(v) = G\left(-\sigma u^j + \sigma f^j\right)$$

(the sum from 1 to (p-1) instead of being from 1 to p is due to the fact that J_N is, here, written with v belonging to E_N).

We shall then define the problem (P_N) by

$$(P_N) \qquad \left\{ \begin{array}{l} \text{Inf} \quad J_N(v) \\ v \in E_N \end{array} \right\}$$

§ 2. Main results.

2.1. The discrete "dual-action principle".

[2.1.1] Proposition 1

Under the hypothesis $[H_o]$, let v be a solution of (P_N), there existsa constant Λ_N such that the piece-wise linear function defined by

$$w_N(t) = \int_o^t v(\tau)\, d\tau + \Lambda_N$$

satisfies the system

$$(H_N) \quad \begin{cases} \dfrac{w(t_{k+1}) - w(t_k)}{t_{k+1} - t_k} = \sigma \nabla H \left(\dfrac{w(t_{k+1}) + w(t_k)}{2} \right) + f(t_k) \\ \forall\, t_k \end{cases}$$

moreover we get

$$\forall\, t \qquad w(t) = w(t+T)$$

in particular

$$w(0) = w(T)$$

Remark

(H_N) is "near" (H) as this can be seen more precisely in 2.2.

[2.1.1] Proof (Sketch)

Step 1

If $u \in \mathbb{R}^{np}$ then the projection, $P_N(u)$, of u on E_N is defined by :

$$P_N(u) = \left(u^j - \frac{1}{p} \sum_{k=1}^{p} u^k\right)_{j=1,p}$$

Step 2

Let d_i^j denote the derivative of J_N with respect to v_i^j , and D_i the projection of d_i on E_N we have :

$$(1) \quad \begin{cases} d^1 = \sigma\left(-\frac{1}{2} \delta^2 v^1 + \delta \nabla G \text{ (at 1)}\right) \\ d^j = \sigma\left(-\frac{1}{2} \delta^2 \left(2 \sum_{j=1}^{p} v^k + v^j\right) + \delta \nabla G \text{ (at j)}\right) \\ \qquad\qquad \text{for } j = 2,p \end{cases}$$

where

$$\nabla G \text{ (at j)} = \nabla G(-\sigma v^j + \sigma f^j)$$

and

$$(2) \quad D^j = d^j - C \qquad j = 1,p$$

where

$$(3) \quad c_i^k = \frac{1}{p} \sum_{j=1}^{p} d_i^j \qquad i = 1,2 \quad k = 1,p$$

(c_i^k is independent of k).

Step 3

If v is a solution of ($\mathcal{P}_N$) we have

$$D^j = 0 \qquad \forall\, j = 1,p$$

and so

$$d^j = c \qquad \forall\, j = 1,p$$

this implies that

$$d^j = d^{j'} \qquad \forall\, j,j'$$

Now extend v to the whole real line, by saying :

$$v(t) = v(\tau) \qquad \text{when } \tau = t - T$$

Step 4

Define Λ_N by

$$\Lambda_N = w(0) = -\frac{1}{2} \delta w^1 + \nabla G \text{ (at 1)}$$

and $w(\cdot)$ by

$$w(t) = \int_0^t v(\tau)\, d\tau + \Lambda_N$$

it is then easy to see :

★ $w(\cdot)$ is piece-wise linear and continuous

★ $w(t) = w(t+T)$

★ $w(t_{k+1}) - w(t_k) = \delta\, v^k = \delta\, v(t_k)$

(remember : $\delta = T/p = t_{k+1} - t_k$)

Step 5

Note that

$$\sigma^{-1} = -\sigma$$

$$w(0) = -\frac{1}{2}\,\sigma\, d^1$$

by Step 3 we have

$$w(0) = -\frac{1}{2}\,\sigma\, d^k$$

$$= -\frac{1}{2}\,\delta\left(2\sum_{j=1}^{k\ 1} v^j + v^k\right) + \nabla G \text{ (at k)}$$

so $w(t_{k+1})$ can be written, using the definition of $w(\cdot)$, as :

$$w(t_{k+1}) = \delta\sum_{j=1}^{k} v^j - \delta\sum_{j=1}^{k-1} v^j - \frac{1}{2}\,\delta\, v^k + \nabla G \text{ (at k)}$$

$$= -\frac{\delta}{2}\, v^k + \nabla G \text{ (at k)}$$

but

$$\delta v^k = w(t_{k+1}) - w(t_k)$$

so

$$\frac{w(t_{k+1}) + w(t_k)}{2} = \nabla G \text{ (at k)}$$

Final step

We have the Fenchel reciprocity formulas [cf. [20]-[21]] that states

$$[a = \nabla G(b)] \Leftrightarrow [b = \nabla H(a)]$$

using that and the previous result we get

$$\frac{w(t_{k+1}) - w(t_k)}{t_{k+1} - t_k} = \sigma \nabla H \left(\frac{w(t_{k+1}) + w(t_k)}{2}\right) + f^k$$

■

2.2. Convergence result.

Take

$$N = 2^{\ell}$$

and define the ℓ-step of the algorithm by solving $(P_{2^{\ell}})$, the solution will be quoted as w_{ℓ}.

[2.2.1] Proposition 2

Under the hypothesis $[H_o]$

(1) w_{ℓ} converges in L^2 , toward $w(\cdot)$

(2) $w(\cdot)$ is a solution of (H) with the boundary constraint [B.C]

Remark

This gives a sense to the remark following proposition 1.

[2.2.2] Proof (Sketch)

Step 1 [see [18]]

(★) The sequence (v_{ℓ}) verifies

$$\forall \ell \qquad \|v_{\ell}\|_{L^2} \leqslant \text{constant}$$

(★) Use the Hilbert-Schmidt property [see [22]] of the operator

$$\alpha \in L^2 \to \int_a^b \alpha(t)\, dt$$

and show the first part of proposition 2.

Step 2 [see [18]]

(★) Define u_ℓ by

$$u_\ell(t_j) = \frac{w_\ell(t_{j+1}) + w_\ell(t_j)}{2}$$

u_ℓ converges in L^2 to $w(\cdot)$.

(★) The sequence

$$\{-\sigma(v_\ell - f_\ell)\}_{\ell \in N} \quad \text{is bounded in } L^2$$

Step 3

The preceding proposition might be written as follows (by remembering that v_ℓ and u_ℓ are step-function)

$$\forall\, t \quad v_\ell(t) = \sigma\, \nabla H\, (u_\ell(v)) + f_\ell(v)$$

or

$$\forall\, t \quad -\sigma(v_\ell(t) - f_\ell(v)) = \nabla H\, (u_\ell(t))$$

and we have

$$\{-\sigma(v_\ell - f_\ell)\}_{\ell \in N} \rightarrow -\sigma(v-f) \quad \text{in } L^2$$

$$u_\ell \rightarrow w \quad \text{in } L^2$$

$$\lim\sup \{(-\sigma(v_\ell - f_\ell))u_\ell\} < -\sigma(v-f)\, u$$

Step 4 (and final one).

Apply the result of J.L. LIONS [23] (p.p. 173-179) that states:

Let π be a maximal monotone operator from its domain into L^2, (b_ℓ) and (c_ℓ) be two sequences in L^2 verifying :

$$(c_\ell) \in \text{Dom}\, \pi$$

$$b_\ell \in \pi\, (c_\ell)$$

$$(c_\ell) \rightarrow \bar{c} \quad \text{(weakly) in } L^2$$

$$b_\ell \rightarrow \bar{b} \quad \text{(weakly) in } L^2$$

$$\lim\sup\, (c_\ell, b_\ell) \leqslant (\bar{c}, \bar{b})$$

then we have

$$\bar{c} \in \mathrm{Dom}\ \pi$$

$$\bar{b} = \pi(\bar{c})$$

using the preceding step we then get :

$$-\sigma(v(t)-f(t)) = \nabla H(w(t)) \qquad \forall\ t$$

so

$$v(t) = \sigma\ \nabla H(w(t)) + f(t) \qquad \forall\ t$$

and

$$w(t) = \int_o^t v(\tau)\ d\tau + \text{constant}$$

$$w(0) = w(T)$$

Q.E.D. ■

§ 3 Numerical application.

3.1. Procedure.

At the step ℓ , if ℓ is large enough, propositions 1 and 2 show us that $w_\ell(\cdot)$ is a good approximation of $w(\cdot)$. So, in general, it is not necessary to iterate by increasing ℓ , it is often sufficient to take it so that $N \approx 100$.

We now have to solve P_N. The principle is quite simple :

- ★ Solve (P_N) by using a gradient - projection method combined with a conjugate-gradient algorithm. See [24] → [27]

- ★ To judge the accuracy, construct $w(\cdot)$, and take $w(0)$ as initial point in a Range-Kutta procedure for solving (H), then compare $w(0)$ with the value in T so obtained.

3.2. Computation of an example.

3.2.1. The example.

$$H(u_1,u_2) = \frac{1}{2} u_1^2 + \frac{1}{4} u_2^4$$

$$G(u_1,u_2) = \frac{1}{2} u_1^2 + \frac{3}{4} u_2^{4/3}$$

$$f(t) = \begin{Bmatrix} \alpha \sin \frac{2\pi t}{T} \\ \alpha \sin \frac{2\pi t}{T} \end{Bmatrix}$$

with other notations we are solving

$$\begin{cases} \ddot{x}(t) - x^3 + \alpha \left(\cos \frac{2\pi t}{T} - \sin \frac{2\pi t}{T}\right) \frac{2\pi}{T} = 0 \\ x(0) = x(T) \end{cases}$$

We took $\alpha = 0{,}1$, $\alpha = 0{,}2$, $\alpha = 0{,}3$, $\alpha > 0{,}3$. The results are not too bad with the two first value of α , but become worse with $\alpha \geqslant 0{,}3$ This might be due to two facts :

- the available conjugate-gradient package are very sensitive to the critical point.
- the hypothesis [4] of [H_o] which bounds α .

3.2.1. Results.

We asked the computer to draw the orbits, that is the points $(w_1(t_j), w_2(t_j))$, when t_j refers to the Runge-Kutta procedure. (This is relevant as long as w(0) characterize the solution, which is our case here).

A^1 , $A^{1'}$, $A^{1''}$ when $\alpha = 0.1$

A/C-1 , A'/C-1 , A"/C-1 when $\alpha = 0.2$.

Figures A 1 and A/C1 show the graph of $w_1(\cdot)$, figures A 1' and A/C1' the corresponding periodic orbit, and figures A 1" and A/C1" a blow-up of the preceding.

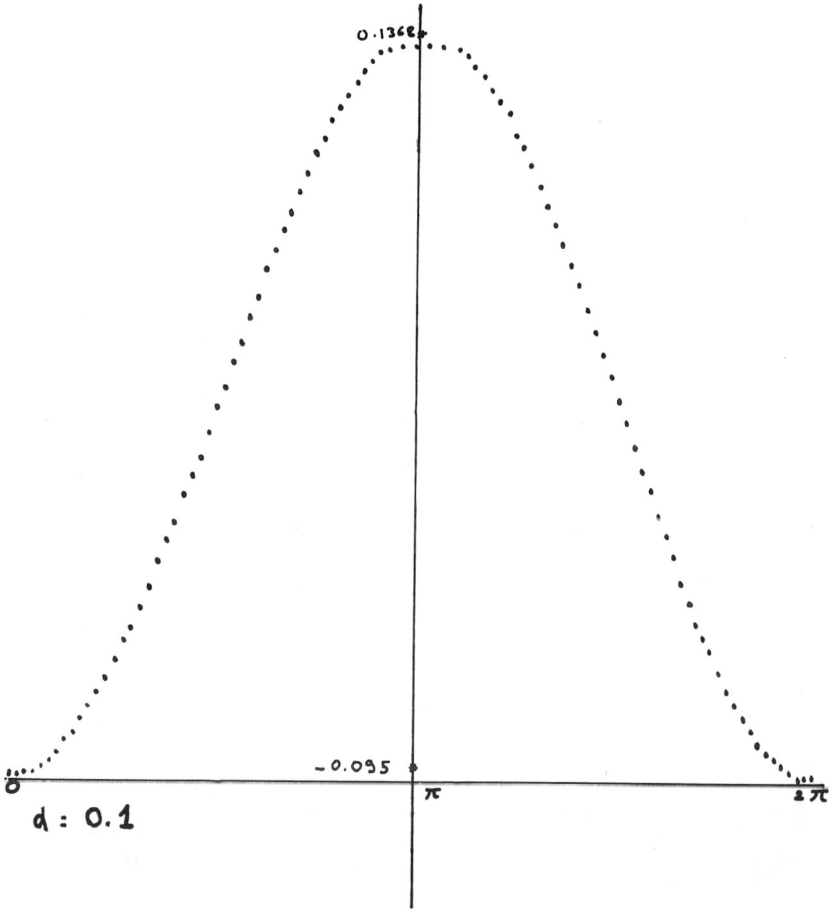

FIG. A1

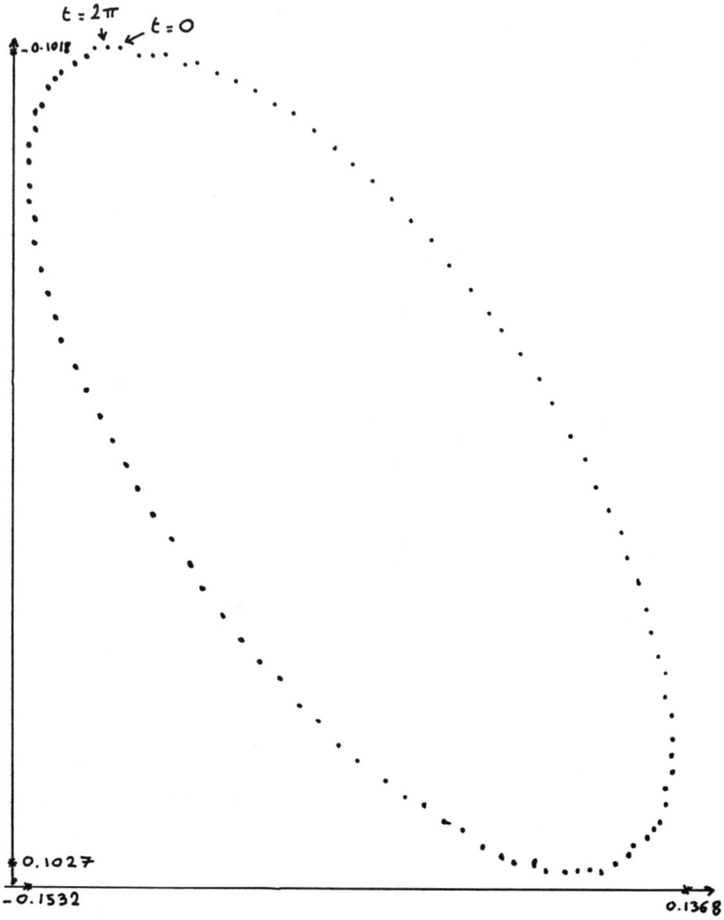

α = 0.1

FIG. A1′

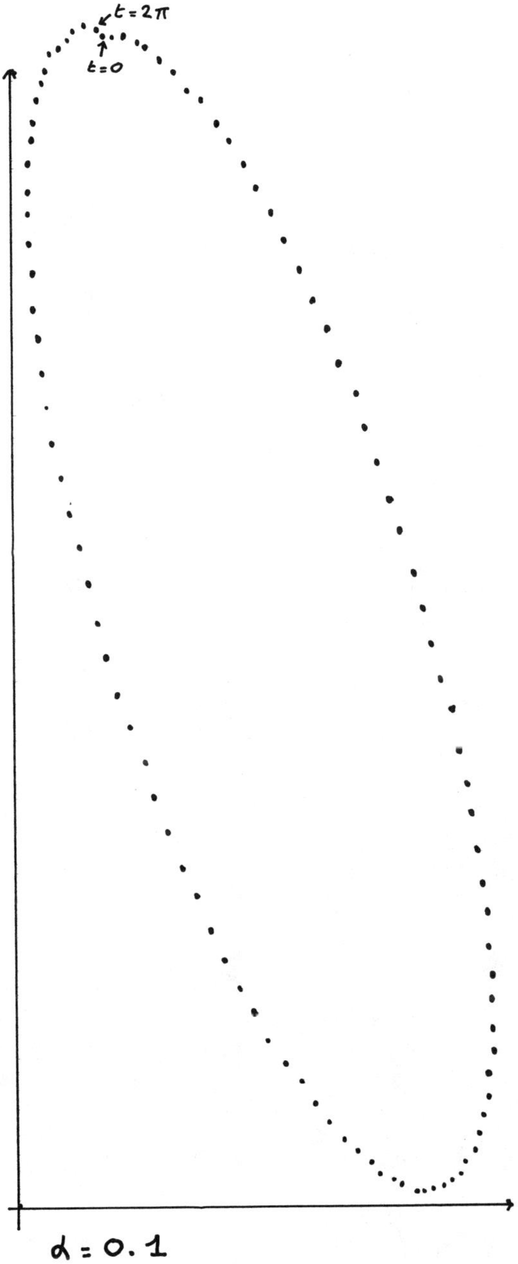

FIG. A1″

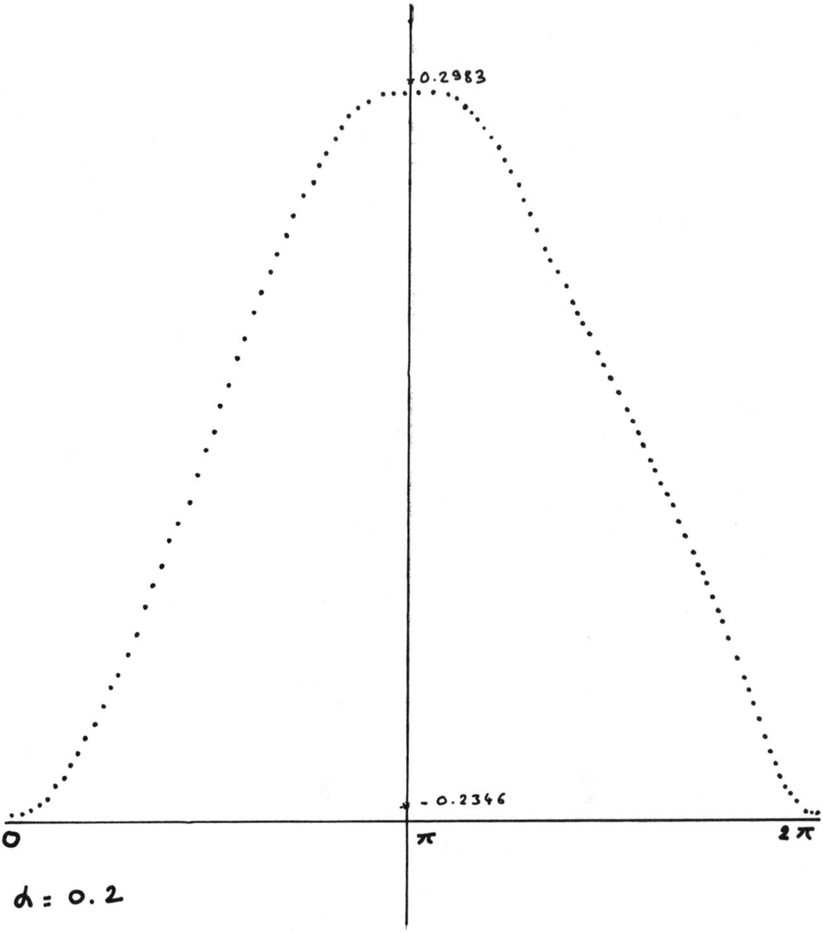

FIG. A/C-1

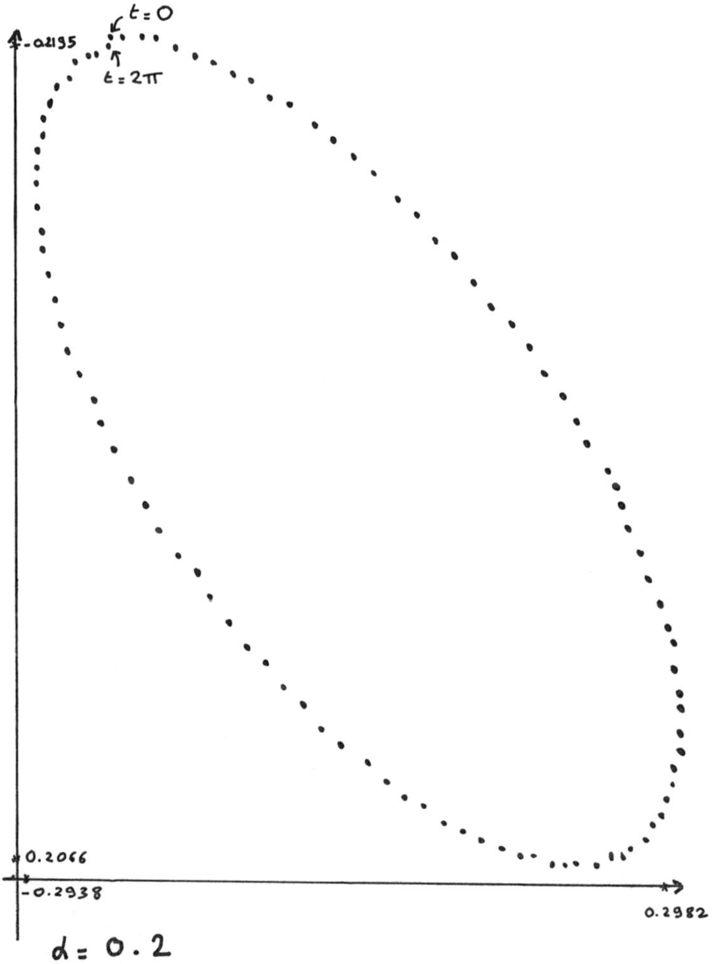

FIG. A'/C-1

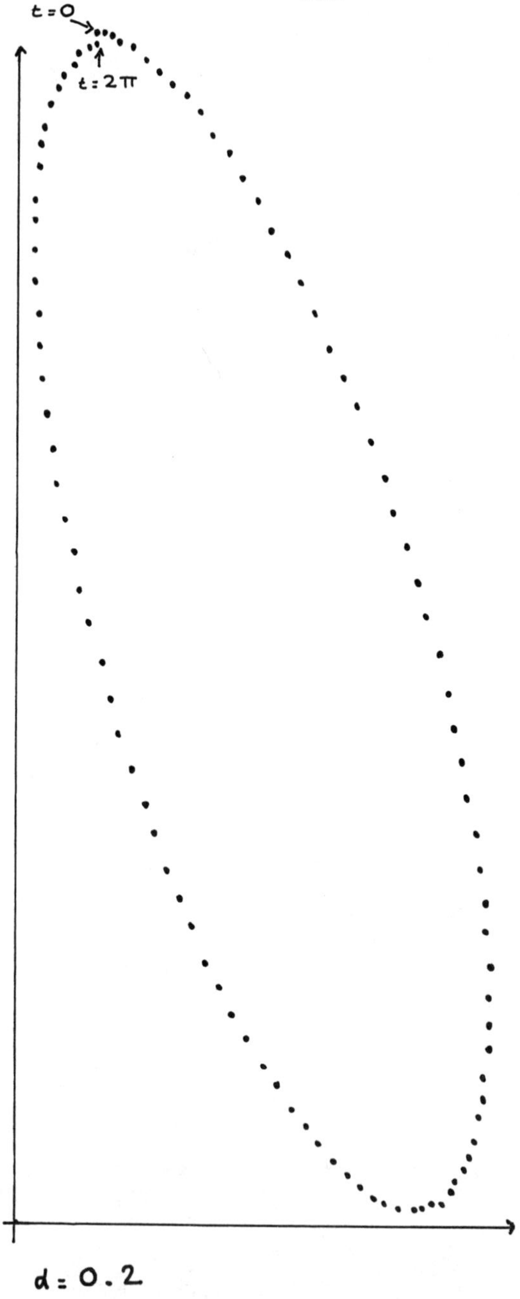

FIG. A''/C-1

BIBLIOGRAPHY

[1] F. Clarke - I. Ekeland : "Hamiltonian trajectories having prescribed minimal period". Comm. Pure and Appl. Math. 33, 1980, p. 103-116.

[2] F. Clarke - I. Ekeland : "Non linear oscillations and boundary value problem for Hamiltonian system". To appear in Archive Rat. Mech. Am.

[3] I. Ekeland "Non linear oscillations of Hamiltonian system. II". Advances in Mathematics, Vol. 7A, 1981, p. 345-360.

[4] I. Ekeland "Non linear oscillations of Hamiltonian system. III". Bull. Soc. Math. France, 109, 1981, p. 297-330.

[5] A. Ambrosetti - P. Rabinowitz : "Dual variationnal methods in critical point theory and applications". J. Funct. Ana. 14 (1973) p. 349-381.

[6] P. Rabinowitz "Periodic solutions of Hamiltonian systems" Comm. Pure and Appl. Math. 31 (1978) p. 157-184.

[7] I. Ekeland "Periodic solutions of Hamiltonian system and a theorem of P. Rabinowitz". J. Diff. Eq. Vol. 34 n° 3, 1979 pp. 523-534.

[8] A. Bahri - H. Beresticky : "Existence d'une infinité de solutions périodiques de certains systèmes hamiltoniens en présence d'un terme de contrainte". C.R. Acad. Sc. Paris tome 292, Série A (1981) 315-318.

[9] Arnol'd "Méthodes Mathématiques de la Mécanique classique". MIR.

[10] Abraham - Marsden : "Foundations of Mechanics". Benjamin / Cumming.

[11] Fermi - Pasta - Ullam : "Studies of non-linear problem". Lectures in Appl. Math. Vol. 15, 1974, p. 143-156.

[12] J. Blot "Recherche de solutions périodiques de systèmes Hamiltoniens perturbés par le théorème des fonctions implicites". Cahier du CEREMADE 8122.

[13] J. Blot Thèse de 3e cycle. Paris IX UNIVERSITY (M.D.) (CEREMADE).

[14] D. Azé Thèse de 3e cycle. Paris IX UNIVERSITY (M.D.) (CEREMADE).

[15] D. Azé - J. Blot : "Les systèmes hamiltoniens et leurs solutions périodiques", CEDIC éditeur, Paris 1982.

[16] I. Ekeland - J.M. Lasry : "On the number of periodic trajectories for an Hamiltonian flow on a convex energy surface". Annals of Math. 112 (1980) p. 283-319.

[17] E. Gaussens "Quelques propriétés topologiques de l'ensemble des solutions périodiques pour des systèmes non linéaires". Cahier du CEREMADE 8101.

[18] E. Gaussens "Recherche numérique des solutions périodiques pour des systèmes non linéaires". Cahier du CEREMADE 8101.

[19] E. Gaussens "Periodic solution in the large for non linear differential systems". To appear in "Numerical functionnal analysis and optimization".

[20] I. Ekeland - R. Temam : "Convex analysis and variationnal problems". North-Holland ELSEVIER 1980.

[21] Rockafellar "Convex analysis". Princeton Univ. Press 1970.

[22] Riesj - Nagy "Leçon d'Analyse fonctionnelle". Gauthiers-Villars.

[23] J.L. Lions "Quelques méthodes de résolution des problèmes aux limites non linéaires". Dunod 1969.

[24] J. Cea "Optimisation : Théorie et Algorithme". Dunod 1971.

[25] J. Abadie "Non linear programming", Wiley 1967.

[26] Auslender - Brodeaux : "Convergence d'un algorithme de Franck-Wolfe appliqué à un problème de contrôle". RIRO n° 7 , 1968.

[27] J.B. Rosen "The gradient projection method for non linear programming". Part I, SIAM 8, pp. 181-217.

N.B.

(I) CEREMADE
Place du Maréchal de Lattre de Tassigny
75775 Paris Cedex 16 FRANCE

(II) Institut de Protection et de Sureté Nucléaire.
Département de Protection.
L.S.E.E.S.
B.P. 6 92260 Fontenay aux Roses
Tel. 654-74-51.